THE SOCIAL IMPACTS OF BIOTECHNOLOGY
AN ANNOTATED BIBLIOGRAPHY OF RECENT WORKS

by

Thomas C. Wiegele
Northern Illinois University

Published by

The Association for Politics and the Life Sciences
Northern Illinois University
DeKalb, Illinois 60115

Table of Contents

SECTION 3: ECONOMIC ASPECTS OF BIOTECHNOLOGY

SECTION 4: INTERNATIONAL BIOTECHNOLOGY

Introduction

As most scholars who conduct interdisciplinary research know, moving into an unfamiliar discipline can be a frightening experience. Not only is one confronted by new terminologies and conceptualizations, but the fear of overlooking a standard piece of literature hangs like the sword of Damocles over the scholar's head. This can lead to a situation in which researchers retreat into the familiar and comfortable confines of their own disciplines, thereby risking little, but surrendering the possibility of innovative connections with potentially helpful bodies of literature.

This bibliography grew out of just such a situation: a frustrating search for materials related to the social impacts of biotechnology. Those materials were (and are) scattered across many disciplines. Information on this topic exists in economics, finance, marketing, sociology, medicine, biology, political science, philosophy, theology, law, and even the literature on higher education. In the best of all possible worlds, one should be able to go to a machine-readable data base which taps all of these disciplines to pull together information on the subject of social impacts of biotechnology. Unfortunately, as far as I know, such a data base does not exist. Thus, this monograph was born to fill a need, at least for the short run until an elaborate data base is constructed.

Readers will note from the subtitle of the volume that this is a bibliography of recent works. Because the subject matter that we are dealing with is expanding so quickly, we decided to concentrate on materials from the 1980s, especially from 1983 to 1986. The

assumption is that this material will incorporate findings from preceding years. This does not mean, and readers can verify this quite quickly, that we have ignored all work prior to 1980. Much of that work is included, but the primary focus remains on the recent past.

It should also be pointed out that the substantive emphasis in this bibliography is on social aspects of biotechnology. This, of course, eliminates the scientific/technical literature on biotechnology, an area of enormous size. This decision had an especially strong impact on the development of the section on ethical aspects of biotechnology. Here the aim was to collate materials on general ethical considerations, rather than on specific ethical factors emerging from individual biotechnologies.

Readers will also note that not a few citations are relevant in several sections of the bibliography. In instances where multiple listings of a single citation occur, the reader is referred to the original annotation.

Numerous individuals participated in this project over the past two years, and they deserve generous thanks for their conscientious service. Two Association for Politics and the Life Sciences Fellows contributed significantly to the digging for source materials and the drafting of the abstracts. Richard Johansen and Karen Kapusinski-Meyer worked diligently in this regard. I am grateful to both of them.

Patricia Finn-Morris applied her skillful editorial eye to numerous drafts of the manuscript and undertook what became a rather complex formatting task. I am appreciative of her efforts.

Lisa Chase, Carolyn Cradduck, J. Patrick Spradlin, and Katharine L. Wiegele assisted in the typing chores. All of them deserve my hearty thanks for a job well done. These individuals, along with Patricia Finn-Morris, also did a considerable amount of proofreading, and they are herewith thanked again.

This project could not have been undertaken without the financial assistance of the Lilly Endowment, Inc. The Endowment provided funds for the support of the Fellows and for the printing of the bibliography. Their generous help as part of a grant to the Association for Politics and the Life Sciences is very much appreciated.

Thomas C. Wiegele

DeKalb, Illinois
June 1986

SECTION 1: REGULATORY ISSUES IN BIOTECHNOLOGY

SECTION 1: REGULATORY ISSUES IN BIOTECHNOLOGY

I. GENERAL

Bazelon, David L. (1983). "Governing Technology: Values, Choices and
    Scientific Progress." Technology in Society 5:15-25.

    This article discusses the judicial review of government
    regulation of public health and safety, and, in the absence of
    such regulation, the role of common law. In relation to
    biotechnology, the author then outlines the ethical and technical
    issues that must be addressed, and the judicial oversight of
    government regulations. He ". . .concludes with comments on the
    public policy process, since the appropriate actions of each of
    the components--government agencies, the courts and the
    public--will affect biotechnology."

Blank, Robert H. (1981). The Political Implications of Human Genetic
    Technology. Boulder, Colo.: Westview Press.

    Blank presents a summary of current facts and discusses future
    prospects of biotechnology in the social and political context of
    the United States. Alternative value frameworks for dealing with
    the issues are presented. Ethical and public policy dimensions
    are examined, leading to the conclusion that these issues will
    ultimately be contested in the political arena.

    _____ (1984). Redefining Human Life: Reproductive Technologies and
    Social Policy. Boulder, Colo.: Westview Press.

    Blank identifies the value context surrounding reproductive
    decision making and describes the state of the art in
    reproductive technologies. The changing legal background is
    discussed and public policy frameworks for present and future
    decision making are examined.

Goggin, Malcolm L. (1984). "The Life Sciences and the Public: Is
    Science Too Important to be Left to the Scientists?" With
    commentaries and author's response. Politics and the Life
    Sciences 3 (1): 28-75.

    In this article and commentary package, the authors examine
    ". . .the arguments for expert self-rule in science and the
    case for more public participation in science policy making."

Green, Harold P. (1983). "The Impact of Biotechnology on Corporate
    Law and Regulation." Technology In Society 5:87-94.

    The author describes the relationship between biotechnology,
    traditional corporate law, and regulation. He then examines the

current regulatory framework, and compares the public policy development with the experience of nuclear power. Public perceptions, moral and ethical issues are also discussed.

Hanson, David J. (1984). "Government Readies Rules for Biotechnology Control." Chemical & Engineering News 62 (33): 34-38.

This article discusses the role, and perceived role, of various government agencies in the development of a regulatory structure adequate for the new technologies. These agencies' lack of experience in the field of biotechnology is also examined.

McGarity, Thomas O. (1985). "Regulating Biotechnology." Issues in Science and Technology 1 (3): 40-56.

McGarity explores the lack of information concerning the consequences that may result from unregulated, or improperly regulated, new biotechnologies. He analyzes the jurisdictions of the various agencies concerned with these technologies and proposes an alternative regulatory framework to the present system.

McGarity, Thomas O., and Karl O. Bayer (1983). "Federal Regulation of Emerging Genetic Technologies." Vanderbilt Law Review 36 (3): 461-540.

This article describes the necessary components of a regulatory system adequately equipped to deal with the new technology available in the field of genetic engineering. It develops this alternative system through a critical analysis of the existing regulatory framework. The authors ". . .recommend that Congress be prepared to enact new legislation if the existing regulatory framework fails to meet the challenge of this exciting new technology."

Moo-young, M., ed. (1985). Comprehensive Biotechnology, Vols. 1-4. New York: Pergamon Press.

Volumes 1 and 2 of this reference set present the unifying, multidisciplinary principles of biotechnology in terms of scientific and engineering fundamentals. The various biotechnological processes, products, and activities that are involved in industry and government are described in volumes 3 and 4. International and national governmental regulations and guidelines on patents, pollution, external aid programs, control of raw materials, and product marketing are discussed. The volumes are indexed, and a detailed glossary of terms is included.

Office of Science and Technology Policy (1984). "Proposal for a Coordinated Framework for Regulation of Biotechnology; Notice." Federal Register. Washington, D.C., December 31.

This proposal is intended to. . ."clarify the policies of the major regulatory agencies that will be involved in reviewing research and products of biotechnology. . .and to explain how the activities of the Federal agencies in biotechnology will be coordinated." The proposal includes an introduction, a seven part regulatory matrix, and statements of proposed policy by the Food and Drug Administration, the Environmental Protection Agency, and the U.S. Department of Agriculture. These agencies propose a regulation framework for the following topics, respectively: biotechnology, microbial products, biotechnology processes and products.

Reilly, Philip (1977). Genetics, Law, and Social Policy. Cambridge, Mass.: Harvard University Press.

Reilly provides a framework within which to consider the following public policy and legal issues: amniocentesis, mass genetic screening, eugenics, genetic counseling, AIH, AID, IVF, cloning, ectogenesis, XYY controversy, and regulation of genetic data banks.

Schmitt, Harrison (1986). "Biotechnology and the Lawmakers." In Joseph G. Perpich (ed.), Biotechnology in Society: Private Initiatives and Public Oversight. Oxford, Great Britain: Pergamon Press, pp. 65-74.

Schmitt, a former senator, explores the effects of the recombinant DNA public policy debates on congressional policy. He examines the relationship between government and the private sector, questioning, for example, what impact governmental regulation will have on scientific and technological innovation. A similar issue, concerning the government's capacity to assess risk and manage it, prompted Schmitt to propose a bill that would allow better definition of the basis for federal regulations. This bill, along with several others, is reviewed. Finally, Schmitt considers the possible effects of biotechnology on other legislative areas.

United States Congress, Office of Technology Assessment (1986). The Regulatory Environment for Science--A Technical Memorandum. Washington, D.C.: U.S. Government Printing Office.

This memorandum ". . .examines the social and legal forces that act to restrict or regulate scientific and engineering research in the United States. . . ." Chapter one presents a summary and subsequent chapters discuss the following: the historical and

political rationales for controls on research; mechanisms for direct and indirect control of research; institutional differences; community control of research; and research policy issues that may warrant future congressional attention. Appendices and references are also provided.

Wade, Nicholas (1984). "Biotechnology and Its Public." Technology in Society 6:17-21.

Wade discusses three key issues of biotechnology: the public debate over hazards, the commercialization of the technology, and the ensuing ethical issues. He then outlines the biologist's lobbying activities against proposed regulation at the federal, state, and local levels. The author concludes that "the more powerful biology becomes, the more its uses and the control of those uses will be debated."

II. HUMAN GENETIC INTERVENTION

A. Genetic Screening and Counseling

Blank, Robert H. (1982). "Public Policy Implications of Human Genetic Technology: Genetic Screening." Journal of Medicine and Philosophy 7 (4): 355-374.

Among the policy considerations discussed is the question of whether genetic screening is a public health matter that needs to be mandated. Blank argues against compulsory mass screening programs except when the disease is easily identified, applicable across groups, and treatable.

Childress, James F., and Kenneth Casebeer (1979). "Public Policy Issues in Genetic Counseling." In Alexander M. Capron, Marc Lappe, Robert F. Murray, Jr., Tabitha M. Powledge, Sumner B. Twiss, and Daniel Bergsma (eds.), Genetic Counseling: Facts, Values, and Norms. New York: Alan R. Liss, Inc.

This chapter provides a framework for examining public policy in relation to genetic counseling. It presents the major arguments that support development of a particular policy to encompass this issue.

Faden, Ruth A., A. Judith Chwalow, Neil A. Holtzman, and Susan D. Horn (1982). "A Survey to Evaluate Parental Consent as Public Policy for Neonatal Screening." American Journal of Public Health 72 (12): 1347-1352.

In 1976, Maryland adopted a regulation requiring parental consent for neonatal screening. This article reviews the results of a study designed to evaluate the effects of this regulation.

Farfel, Mark R., and Neil A. Holtzman (1984). "Education, Consent, and Counseling in Sickle Cell Screening Programs: Report of a Survey." American Journal of Public Health 74 (4): 373-375.

In 1980, a survey of sickle cell screening programs was conducted in order to assess their compliance with Maryland regulations. The survey reported the following: approximately 25 percent were screened without informed consent; many facilities provided inadequate counseling and education; and facilities providing screening services exclusively were the most compliant with the state regulations.

Kodama, Kyoko, Toshikazu Nakata, Jyoji Ishii, Kazunori Mitani, Hidehiko Tsunooka, Akira Masaoka, and Mitsuko Aoyama (1985). "VMA Mass Screening Program of Neuroblastoma for Infants in Nagoya City, Japan." American Journal of Public Health 75 (2): 173-175.

This article discusses the VMA mass screening program for detection of neuroblastoma, a common tumor of childhood, in Nagoya City, Japan. The methods used and results attained are reported.

Lappe, Marc (1979). "Genetic Screening." In Y. Edward Hsia, Kurt Hirschhorn, Ruth L. Silverberg, and Lynn Godmilow (eds.), Counseling in Genetics. New York: Alan R. Liss, Inc., pp. 295-309.

Lappe describes the three basic groups of genetic screening programs: presymptomatic, parental, and research.

III. PRENATAL ISSUES

A. Reproductive Technologies

Bonnicksen, Andrea L. (1984). "In Vitro Fertilization, Artificial Insemination, and Individual Rights: A Review of Policy." Paper presented at the annual meeting of the American Political Science Association, Washington, D.C., August 30-September 2.

It is estimated that one in six married couples is infertile. Alternative reproductive technologies offer them the opportunity of biological parenthood. Bonnicksen assesses the effect of case law and statutes on biopolicy. The author concludes that individual rights are not threatened by the technology.

Department of Health & Social Security (1984). Report of the
    Committee of Inquiry into Human Fertilisation and Embryology.
    London: Her Majesty's Stationary Office.

    This report examines the social, ethical, and legal implications
    of recent and potential developments for the alleviation of
    infertility. The techniques of these developments are described.
    Chapter 13 discusses regulating infertility services and
    research. A list of recommendations and expressions of dissent
    by committee members are also presented.

Grobstein, Clifford, Michael Flower, and John Mendeloff (1983).
    "External Human Fertilization: An Evaluation of Policy."
    Science 222 (4620): 127-133.

    This article presents the current status of public policy in the
    area of in vitro fertilization. It also provides a discussion of
    the supply, demand, and supply of services concerned with this
    technology.

Murray, Robert F. (1981). "In Vitro Fertilization and Embryo
    Transfer: The Process of Making Public Policy." In Helen B.
    Holmes, Betty B. Hoskins, and Michael Gross (eds.), The
    Custom-Made Child? Clifton, N.J.: The Humana Press, Inc.,
    269-274.

    Murray provides a brief outline describing components of
    procedures used in in vitro fertilization and embryo transfer.
    He discusses the underlying reasons for the recommendations made
    by the Ethics Advisory Board to the Secretary of the Department
    of Health, Education, and Welfare concerning these procedures.

B. Fetal Research

Baron, Charles H. (1985). "Fetal Research: The Question in the
    States." The Hastings Center Report 15 (2): 12-16.

    At the federal level, regulation of fetal research is limited to
    "all research involving human subjects conducted by the
    Department of Health and Human Services or funded in whole or in
    part by a Department grant, contract, cooperative agreement, or
    fellowship." These regulations are not applicable to research
    not funded or conducted by the department. This article examines
    the various state regulations, and the variability of these
    regulations. The author then discusses fetal research in
    relation to rule making in a pluralist society. He concludes
    that state regulations are far from perfect. They allow, however,
    for refinements through argument, experience, and negotiation.

Fletcher, John C., and Joseph D. Schulman (1985). "Fetal Research: The State of the Question." The Hastings Center Report 15 (2): 6-12.

Fletcher and Schulman argue that, since 1980, federal fetal research policy has been severely impaired by the lack of an Ethics Advisory Board (EAB). In their examination of fetal research, the authors provide discussions on the early debates, the EAB, present research, federal regulations, and the consequences of inaction. The authors state, "our goal is the development of a set of public policies, ethically and institutionally coherent, to protect those toward whom research may be directed."

Levine, Robert J. (1981). Ethics and Regulation of Clinical Research. Baltimore, Md.: Urban & Schwarzenberg.

This book presents a survey of the ethical and legal duties of clinical researchers. The author is particularly concerned with federal regulations for the protection of human research subjects. Issues discussed include the following: basic concepts and definitions; ethical norms and procedures; selection of subjects; informed consent; children; the mentally ill; prisoners; the fetus; and the Institutional Review Board. Appendices are also presented which include the rules and regulations of the Department of Health and Human Services and the Food and Drug Administration.

C. Sex Preselection

Wiegele, Thomas C. (1985). "The Biotechnology of Sex Preselection: Social Issues in a Public Policy Context." Policy Studies Review 4 (3): 445-460.

Wiegele discusses several methods of sex selection, the likely techniques of sex preselection, and the social and political implications of a potential sex preselection technology. "Confronted with a major problem of public policy," the author states, "government would have five broad major choices: (1) prohibit, (2) regulate, (3) encourage, (4) mandate, or (5) take no action." Each policy option is examined.

D. Abortion

Legge, Jerome S., Jr. (1983). "The Unintended Consequences of Policy Change: The Effect of a Restrictive Abortion Policy." Administration & Society 15 (2): 243-256.

In order effectively to analyze the implications of proposed policy change in the United States concerning abortion, the author presents a study of nations that have moved from a

"liberal" abortion policy to a more restrictive one. The
countries that most typify this movement are the States of
Eastern Europe. The study examines the maternal health
consequences of such a policy change. It concludes that, in the
case of Romania, such an abrupt policy reversal may have
disastrous effects upon maternal health.

IV. ISSUES WITHIN THE LIFE CYCLE

A. Organ Transplantation

Annas, George J. (1985). "Regulating the Introduction of Heart and
    Liver Transplantation." Public Health and the Law 75 (1): 93-95.

    This is a summary of a report, and recommendations made by the
    Massachusetts Task Force on organ transplantation. The task
    force was created in order to examine the social issues concerned
    with transplantation technology. The author notes, "the
    Massachusetts experience and the Task Force recommendations are
    of special interest because Massachusetts is one of only a few
    states to use the determination of need (DON) process to regulate
    organ transplantation."

Gorovitz, Samuel (1984). "Against Selling Bodily Parts." QQ-Report
    from the Center for Philosophy and Public Policy 4 (2): 9-12.

    This article presents an ethical and policy argument against the
    use of the free market system to establish a commercial market in
    transplantable organs.

Reiss, John B., John Burckhardt, and Fred Hellinger (1982). "Costs
    and Regulation of New Medical Technologies: Heart Transplants as
    a Case Study." In Barbara J. McNeil, and Ernest G. Cravalho
    (eds.), Critical Issues in Medical Technology. Boston, Mass.:
    Auburn House Publishing Company.

    The authors examine policy and regulatory issues in context with
    government payments for new medical procedures and technologies.
    A case study of heart transplantations provides the research
    focus. The costs and benefits in relation to social and economic
    factors are discussed. Potential federal program costs and
    associated benefits are also considered. The authors then
    address the ". . .social, ethical, and legal concerns about the
    regulatory implications of funding heart transplantation."

Simmons, Roberta G., and Susan Klein Marine (1984). "The Regulation of High Cost Technology Medicine: The Case of Dialysis and Transplantation in the United Kingdom." Journal of Health and Social Behavior 25 (3): 320-334.

The structure and operational policies of the British National Health Services (NHS), with respect to how they affect the allocation of high cost medical technology, are discussed. The research focus of this article is a case study of the technologies of dialysis and kidney transplantation. Comparisons to the U.S. are presented when relevant. The authors note, "when governments intervene to alleviate. . .financial pressures and make the delivery of care more equitable, regulation of cost may—perhaps inevitably—become an issue."

Wehr, Elizabeth (1984). "Special Report: The Politics of Organ Transplants." Congressional Quarterly 42 (8): 453-458.

This article discusses the legislative bill on organ transplantation proposed by Representative Albert Gore, Jr. Reactions to this proposed legislation and its possible effects are presented.

B. Human Experimentation

Culliton, Barbara J. (1985). "News and Comment: Gene Therapy-Research in Public." Science 227:493-496

According to Culliton, the current trend in gene therapy research will soon lead to human experimentation. This experimentation will be subject to government policy requiring regulation and public debate.

Levine, Robert J. (1981). Ethics and Regulation of Clinical Research. Baltimore, Md.: Urban & Schwarzenberg.

See Regulation, III. Prenatal Issues, B. Fetal Research, above.

Sass, Hans-Martin (1983). "Reichsrundschreiben 1931: Pre-Nuremberg German Regulations Concerning New Therapy and Human Experimentation." Journal of Medicine and Philosophy 8 (2): 99-112.

Sass examines the regulation of human experimentation during the Third Reich.

Veatch, Robert M., and Herman S. Wigodsky (1981). "Two Views of the
New Research Regulations." The Hastings Center Report 11 (3):
9-14.

These authors examine the regulations governing research
involving human subjects. Veatch, in "Protecting Human Subjects:
The Federal Government Steps Back," concludes some improvements
have been made. He is, however, concerned about the lessened
degree of federal protection. Wigodsky, in "New Regulations, New
Responsibilities for Institutions," notes "a considerable
achievement" in the new regulations, but raises some questions
concerning their implementation.

V. BIOPOLICY

Blank, Robert H. (1982). "Biopolicy: A Restatement of Its Role in
Politics and the Life Sciences." With commentaries and author's
response. Politics and the Life Sciences 1 (1): 38-51.

In this article and commentary package, Blank argues that
"students of politics and the life sciences have consequential
contributions to make to biopolicy." He then discusses areas in
biopolicy of substantive concern, and the contributions of
politics and the life sciences in this area. Issues of
biomedical policy formulation are also presented.

_____ (1986). "Public Administration and Biopolicy." In Elliott
White (ed.), Biology and Bureaucracy. Lanham, Md.: University
Press of America.

Blank defines the role of the political scientist in resolving
public policy questions created by biotechnology. He identifies
individual, group, and global areas of substantive concern for
biopolicy, and questions the ability of existing institutions to
deal with issues created by the technology. He states that the
role of the political scientist is to define the problem and
clarify the value issues, in order to make biopolicy.

Funke, Odelia (1985). "Biopolitics and Public Policy: Controlling
Biotechnology." PS 18 (1): 69-77.

The author is concerned with the need for more biopolitical
research for the formulation of public policy, particularly
regulatory policy, in relation to biotechnology. She examines
the complex issues confronting federal regulatory agencies, and
their ability to deal with these issues. Funke concludes that
". . .since there are complex scientific as well as social issues
involved in these emerging technologies, a full understanding
requires some knowledge in both of these fields. Biopolitics
provides this perspective."

Zilinskas, Raymond A. (1983). "New Biotechnology: Potential Problems Likely Promises." With commentaries and author's response. Politics and the Life Sciences 2 (1): 42-75.

The author discusses the role of the social scientist in bridging the gap between science and politics through examining the policy implications of new biotechnology. The use of international organizations in technology transfer is also considered. Commentaries expand on arguments made.

VI. AGRICULTURE AND ENVIRONMENT

Alexander, Martin (1985). "Ecological Consequences: Reducing the Uncertainties." Issues in Science and Technology 1 (3): 57-68.

This article is concerned with the potential damage that genetically engineered organisms may have on the ecosystem to which they are introduced. The author provides a set of guidelines that, he argues, should be followed during research and testing to determine whether an organism will be harmful to the environment. "Compliance with the appropriate regulations," he states, "can significantly reduce the possibility of an ecological upset."

Brill, Winston J. (1985). "Safety Concerns and Genetic Engineering in Agriculture." Science 227 (4685): 381-384.

Various regulations are being considered by federal agencies for the protection of the public from environmental and health problems that might arise from the release of genetically engineered organisms. In this article, the author examines the safety issues concerned with the use of genetically engineered plants and microorganisms. He discusses the need for field testing, and states that "regulations governing release of genetically engineered organisms should be based on scientific experience and informed debate of the issues."

United States Congress. House Subcommittee on Science, Research, and Technology of the Committee on Science and Technology (1983). Environmental Implications of Genetic Engineering. 98th Congress, 1st session. Publication number 98-36. Washington, D.C.: U.S. Government Printing Office.

Transcript of a congressional hearing that considered whether deliberate release of novel organisms (plants and microbes) would damage the ecosystem. For agriculture this includes the use of genetic biotechnology for plant growth regulation, biocontrol of pests, and frost resistance. For other areas it covers the use of engineered microbes to break down toxic waste and produce new forms of energy. The answer regarding environmental damage falls

somewhere between "don't know" and "some may." These possibilities are weighted against benefits to be derived, in a context that asks what is the role of the federal government for policy and regulation.

United States Congress, Office of Technology Assessment (1981). _Impacts of Applied Genetics: Micro-Organisms, Plants, and Animals_. Washington, D.C.: U.S. Government Printing Office.

A book that examines applications of 'old' and 'new' genetic technologies to microbes, plants, and animals. Chapters 1, 11, 12, and 13 are of particular interest. The first is a summary chapter, and the subsequent chapters discuss the following: regulation of genetic engineering; patenting living organisms; genetics and society. Issues and options for the Congress are also identified and discussed. The book has excellent appendices that include: case studies, reproductive technologies, history of the rDNA debate, constitutional constraints on regulation, and international guidelines.

_____ (1982). _Genetic Technology: A New Frontier_. Boulder, Colo.: Westview Press.

This is a reprint of the government report, _Impacts of Applied Genetics: Micro-Organisms, Plants, and Animals_.

VII. INDUSTRIAL

Hanson, David J. (1985). "Government, Industry Officials Discuss Biotechnology Public Policy." _Chemical & Engineering News_ 63 (5): 21-23.

Hanson presents a report of the conference held at the Brookings Institution in Washington, D.C., where government and industry officials discussed public policy perspectives on biotechnology. The article states that two important points were made. First, "overreacting to the perceived dangers of biotechnology. . .will hurt research and its application." And, "the public's perceptions will determine how these technologies are regulated."

Hardy, Ralph W. F., and David J. Glass (1985). "Our Investment: What is at Stake?" _Issues in Science and Technology_ 1 (3): 69-82.

This article discounts the concerns of those who argue that the scientific community has not investigated the potential dangers new genetic technologies may pose to human health. The authors are particularly concerned with the effect of overly stringent regulations on the development of this important U.S. industry.

14

Johnson, Irving S. (1986). "National Policy and Biotechnology in the
    United States." In Joseph G. Perpich (ed.), <u>Biotechnology</u> <u>in</u>
    <u>Society</u>: <u>Private</u> <u>Initiatives</u> <u>and</u> <u>Public</u> <u>Oversight</u>. Oxford,
    Great Britain: Pergamon Press, pp. 213-215.

Johnson asserts that the U.S. ". . .appears to be floundering
around for a national policy on biotechnology." He then compares
the U.S. efforts with those of Japan and the United Kingdom.
Recommendations concerning regulation of biotechnology are
presented, including a recommendation for increasing the role of
industry in the policy process.

Korwek, Edward L. (1982). "FDA Regulation of Biotechnology as a New
    Method of Manufacture." <u>Food</u> <u>Drug</u> <u>Cosmetic</u> <u>Law</u> <u>Journal</u> 37 (3):
    289-309.

Korwek argues that manufacturers of biotechnology-produced
articles should develop a set of standards for assuring the
purity of these articles. If this is accomplished, the author
contends that the FDA should not impose any unique set of
regulations on items produced by the new technologies.

Mitman, Hank (1986). "Biotechnology--Export Controls." In Joseph G.
    Perpich (ed.), <u>Biotechnology</u> <u>in</u> <u>Society</u>: <u>Private</u> <u>Initiatives</u> <u>and</u>
    <u>Public</u> <u>Oversight</u>. Oxford, Great Britain: Pergamon Press, pp.
    209-212.

The current situation and the future direction of the Office of
Export Administration (OEA) is the focus of this article. Also,
recommendations made by the Office of Science and Technology
Policy and the Department of Commerce concerning controls and
regulation of biotechnology are discussed.

Moo-young, M., ed. (1985). <u>Comprehensive</u> <u>Biotechnology</u>, Vols. 1-4.
    New York: Pergamon Press.

See Regulation, I. General, above.

Perpich, Joseph G. (1983). "Genetic Engineering and Related
    Biotechnologies: Scientific Progress and Public Policy."
    <u>Technology</u> <u>in</u> <u>Society</u> 5:27-49.

Perpich examines the NIH's impact on the development of
federal policies in the field of biotechnology. He outlines
". . .research activities and current federal policies and
programs that might spur the goals of the biotechnology
industry." He concludes with a proposal for increasing
cooperation between government, universities, and industries in
biotechnology programs.

Wade, Nicholas (1984). "Biotechnology and Its Public." Technology in Society 6:17-21.

See Regulation, I. General, above.

VIII. INTERNATIONAL

Cassell, Paul G. (1983). "Establishing Violations of International Law: 'Yellow Rain' and the Treaties Regulating Chemical and Biological Warfare." Stanford Law Review 35 (2): 259-295.

This article is concerned with the international community's efforts to regulate the development and use of chemical and biological weapons. The author argues that the 1925 Geneva Protocol and the 1972 Biological Weapons Convention are ill equipped to provide the means for investigating violations. He concludes that a new organization should be created.

Johnson, Irving S. (1986). "National Policy and Biotechnology in the United States." In Joseph G. Perpich (ed.), Biotechnology in Society: Private Initiatives and Public Oversight. Oxford, Great Britain: Pergamon Press, pp. 213-215.

See Regulation, VII. Industrial, above.

Kodama, Kyoko, Toshikazu Nakata, Jyoji Ishii, Kazunori Mitani, Hidehiko Tsunooka, Akira Masaoka, and Mitsuko Aoyama (1985). "VMA Mass Screening Program of Neuroblastoma for Infants in Nagoya City, Japan." American Journal of Public Health 75 (2): 173-175.

See Regulation, II. Human Genetic Intervention, A. Genetic Screening and Counseling, above.

Krimsky, Sheldon (1984). "Regulation Policies on Biotechnology in Canada." Report prepared for the Science Council of Canada, October.

Various sections discuss the following: commercial, agricultural, and environmental dimensions of biotechnology; the legal and administrative structure pertaining to the regulation of biotechnology; U.S. regulatory responses; and regulatory policy considerations for Canadian biotechnology.

Mitman, Hank (1986). "Biotechnology--Export Controls." In Joseph G. Perpich (ed.), Biotechnology in Society: Private Initiatives and Public Oversight. Oxford, Great Britain: Pergamon Press, pp. 209-212.

See Regulation, VII. Industrial, above.

Moo-young, M., ed. (1985). Comprehensive Biotechnology. Vols. 1-4.
New York: Pergamon Press.

See Regulation, I. General, above.

Perpich, Joseph G. (1983). "Genetic Engineering and Related
Biotechnologies: Scientific Progress and Public Policy."
Technology in Society 5:27-49.

See Regulation, VII. Industrial, above.

Rogers, Michael D. (1982). "The Japanese Government's Role in
Biotechnology R & D." Chemistry and Industry 15:533-537.

This article provides a description of the Science Council of
Japan (SCJ) and its advisory relationship to the government.
Further, the author examines how SCJ recommendations are filtered
through the government structure.

Somjee, A.H. (1982). "The Techno-Managerial and Politico-Managerial
Classes in a Milk Cooperative of India." Journal of Asian and
African Studies 17:122-134.

In the absence of bureaucratic intervention, technocrats are able
to build a modern institutional network that achieves rational,
scientific goals.

United States Congress, Office of Technology Assessment (1984).
Commercial Biotechnology: An International Analysis.
OTA-BA-218. Washington, D.C.: U.S. Government Printing Office.

Topics covered include the industrial use of rDNA, cell fusion
and novel bioprocessing techniques for agriculture,
pharmaceuticals, energy, and bioelectronics. The competitive
position of the United States in this field with that of five
other states is assessed. Separate chapters discuss the relative
importance of ten factors that determine a state's competitive
position. Policy options are identified with respect to these
factors. Chapters 15, 20, and 21 are of special interest and
discuss the following: health, safety, and environmental
regulation; targeting politics in biotechnology; and public
perception. The book has a summary, a description of
technologies, and excellent appendices.

Wolstenholme, Gordon (1984). "Public Confidence in Scientific
Research: The British Response to Genetic Engineering."
Technology in Society 6:9-16.

The author, former chairman of the Genetic Manipulation Advisory
Group (GMAG), outlines the United Kingdom's guidelines for
recombinant DNA research. He compares the work of the NIH with

that of the GMAG, and its collaboration with the NIH and similar
scientific bodies in Europe. Past efforts to create common
standards within the international community are examined. He
concludes that an advisory board, similar to the GMAG, should be
created at the international level.

SECTION 2: LEGAL ASPECTS OF BIOTECHNOLOGY

SECTION 2:   LEGAL ASPECTS OF BIOTECHNOLOGY

I. GENERAL

Blank, Robert H. (1984). "The Changing Legal Framework of
    Reproductive Choice." In Redefining Human Life: Reproductive
    Technologies and Social Policy. Boulder, Colo.: Westview Press,
    pp. 73-112.

    Blank discusses the response of the law to the new reproductive
    technologies. This includes the following topics: abortion,
    sterilization, genetic screening, artificial insemination,
    wrongful life.

Lygre, David G. (1979). Life Manipulation. New York: Walker and
    Company.

    This book describes the technology and then discusses the social,
    ethical, and legal issues raised by artificial insemination, IVF,
    embryo banks, artificial wombs, cloning, genetic therapy, genetic
    counseling, prenatal screening, abortion, organ transplantation,
    psychosurgery, and freedom of scientific inquiry.

Milunsky, Aubrey, and George J. Annas, eds. (1976). Genetics and the
    Law. New York: Plenum Press.

    An overview of legal issues raised by biotechnology: fetal
    viability, prenatal diagnosis, genetic screening, genetic
    counseling, informed consent, AID, sterilization, IVF, cloning,
    eugenics, state channeling of gene flow.

Reilly, Philip (1977). Genetics, Law, and Social Policy. Cambridge,
    Mass.: Harvard University Press.

    Reilly provides a framework within which to consider the
    following legal and public policy issues: amniocentesis, mass
    genetic screening, eugenics, genetic counseling, AIH, AID, IVF,
    cloning, ectogenesis, XYY controversy, regulation of genetic data
    banks.

Uzych, Leo (1986). "Genetic Testing and Exclusionary Practices in the
    Workplace." Journal of Public Health Policy 7 (1): 37-57.

    Public policy issues associated with genetic testing of workers
    and policies that exclude women from jobs that pose possible
    exposure to reproductive hazards is the focus of this article.
    Topics discussed include the following: legal challenges to

exclusionary practices; legislative action; major legal and
contentious public policy-related issues; and deficiencies in
legal protection. A discussion concerning the need for
guidelines is also presented.

II. HUMAN GENETIC INTERVENTION

A. Genetic Counseling

Reilly, Philip (1979). "Genetic Counseling: A Legal Perspective." In
    Y. Edward Hsia, Kurt Hirschhorn, Ruth L. Silverberg, Lynn
    Godmilow (eds.), Counseling in Genetics. New York, N.Y.: Alan
    R. Liss, Inc., pp. 311-328.

    The legal nature of the standard of care, the use and misuse of
    confidential data, and the scope of liability for mistakes is
    discussed in this book.

B. Wrongful Life

Annas, George J. (1979). "Medical Paternity and 'Wrongful Life.'" The
    Hastings Center Report 9 (3): 15-17.

    Annas discusses litigation resulting from advances in
    biotechnology, specifically, the prenatal diagnosis of birth
    defects.

_____ (1981). "Righting the Wrong of 'Wrongful Life'." The Hastings
    Center Report 11 (1): 8-9.

    A further discussion of the legal concept of "wrongful life" in
    light of a 1980 California court decision.

Furrow, Barry R. (1983). "Impaired Children and Tort Remedies: The
    Emergence of a Consensus." Law, Medicine and Health Care 11 (4):
    148-154.

    In his commentary on suits for wrongful life, Furrow includes a
    discussion of the Harbeson case.

Harbeson v. Parke-Davis, Inc. (1983). Wash., 656 P. 2d 483.

    In this court case the Supreme Court of Washington State held
    that parents can maintain action for wrongful birth and children
    can maintain action for wrongful life. A suit for wrongful life
    may be brought on behalf of a defective infant against a
    physician that failed to give information about a possible
    defect. Liability is created by diminished quality of life.

21

Teff, Harvey (1985). "The Action for 'Wrongful Life' in England and the United States." <u>International</u> <u>and</u> <u>Comparative</u> <u>Law</u> <u>Quarterly</u> 34 (3): 423-441.

> Issues concerning the action for "wrongful life" discussed in this article include past cases, the analogy of "wrongful birth," and objections to such action based on technical and policy considerations. The author argues that ". . .at least in cases of severe congenital deformity, the affected child has a morally compelling claim to redress, fully compatible with common law principles and the requirements of public policy."

C. Eugenics

Eisenstadt v. Baird (1972). 405 U.S. 438.

> In this case, the court recognized the right of the individual to be free from unwarranted governmental intrusion into the decision as to whether to bear a child.

Weigel, Charles L. II, and Stephen E. Tinkler (1973). "Eugenics and Law's Obligation to Man." <u>South</u> <u>Texas</u> <u>Law</u> <u>Journal</u> 14:361-391.

> Positive eugenics involves restructuring chromosomes either to avoid a deleterious effect or to achieve a positive one. Negative eugenics is the use of genetic counseling or screening for the same purpose. This article surveys the history and present day status of eugenics, and then examines legal action and reaction to the issues raised.

III. PRENATAL ISSUES

A. Reproductive Technologies

Bonnicksen, Andrea L. (1984). "<u>In</u> <u>Vitro</u> Fertilization, Artificial Insemination, and Individual Rights: A Review of Policy." Paper presented at the annual meeting of the American Political Science Association, Washington, D.C., August 30-September 2.

> See Regulation, III. Prenatal Issues, Reproductive Technologies, above.

Smith, Lucinda Ann (1979). "Artificial Insemination: Disclosure Issues." <u>Columbia</u> <u>Human</u> <u>Rights</u> <u>Law</u> <u>Review</u> 11 (1): 87-101.

> The issue of disclosure--whether a child conceived through artificial insemination has a right to know the identity of his or her biological father--is discussed.

22

Steeves, Sharon M. (1979). "Artificial Human Reproduction: Legal
     Problems Presented by the Test Tube Baby." Emory Law Journal 28
     (4): 1045-1079.

     Steeves considers questions raised by reproduction technology:
     What are the legal rights of human offspring conceived by other
     than natural means? Does the state have an interest in
     regulating reproduction?

B. Abortion

Blank, Robert H. (1984). "Judicial Decision Making and Biological
     Fact: Roe v. Wade and the Unresolved Question of Fetal
     Viability." The Western Political Quarterly 37 (4): 584-602.

     Blank discusses the inability of the courts to deal effectively
     with legal issues raised by new biotechnology. He specifically
     cites the case of Roe v. Wade, stating that ". . .advances in
     prenatal intervention and in neonatal intensive care. . .are
     altering the biological bases upon which the court built its
     decision."

Goldstein, Mark A. (1978). "Choice Rights and Abortion: The Begetting
     Choice Right and State Obstacles to Choice in Light of Artificial
     Womb Technology." Southern California Law Review 51 (5):
     877-921.

     According to Goldstein, artificial womb technology, which allows
     the fetus to exist independently of the mother before viability,
     raises questions as to whether the state may proscribe abortion
     earlier than viability.

Roe v. Wade (1973). 410 U.S. 113.

     A landmark case which decided under what circumstances a woman
     may have a legal abortion.

Fletcher, John C., et al. (1980). "Prenatal Diagnosis for Sex
     Choice." Article and Commentaries. The Hastings Center Report
     10 (1): 15-20.

     In this article/commentary package, the issues surrounding the
     use of abortion as a method of gender selection are debated.

C. Personhood

Lenow, Jeffrey L. (1983). "The Fetus as a Patient: Emerging Rights
     as a Person?" American Journal of Law and Medicine 9 (1): 1-29.

     This article identifies medicolegal conflicts that arise as fetal
     surgery becomes established.

23

Milby, T. H. (1983). "The New Biology and the Question of Personhood:
     Implications for Abortion." American Journal of Law and Medicine
     9 (1): 31-41.

     The article states that understanding the phenomena of chimerism,
     cloning, and parthenogenesis may relieve moral ambiguity about
     abortion.

IV. ISSUES WITHIN THE LIFE CYCLE

A. Organ Transplantation

Mack, Eric (1980). "Bad Samaritanism and the Causation of Harm."
     Philosophy and Public Affairs 9 (3): 230-259.

     This article raises social science value issues relevant to organ
     transplantation and organ banks. Individuals do not have a legal
     duty to save the life of another; Mack discusses Bad Samaritan
     laws which would create a legal duty. The discussion covers both
     legal and value issues.

Norrie, Kenneth McK. (1985). "Human Tissue Transplants: Legal
     Liability in Different Jurisdictions." International and
     Comparative Law Quarterly 34 (3): 442-469.

     The author examines the legal liability of human tissue
     transplants in the British, American, and other legal systems.
     Differences that exist between transplants from living donors
     and transplants from deceased donors provide the focus for the
     discussion. He then discusses the problem of scarce resources
     and concludes, "The only way in the end to prevent such
     problems. . .is to encourage to a far greater degree the
     donation of organs."

V. DEATH: RELATED ISSUES

A. Euthanasia

Annas, George J. (1980). "Quinlan, Saikewicz, and Now Brother Fox."
     The Hastings Center Report 10 (3): 20-21.

     Annas makes the point that issues such as termination of medical
     treatment must be decided on a case by case basis, due to lack of
     legal precedent and accepted social standards.

24

_____ (1980). "Quality of Life in the Courts: Early Spring in
Fantasyland." The Hastings Center Report 10 (4): 9-10.

An analysis of court cases dealing with the concept of "quality
of life," and the medical, legal, and ethical issues raised by
the decisions.

_____ (1983). "Nonfeeding: Lawful Killing in California, Homicide in
New Jersey." The Hastings Center Report 13 (6): 19-20.

The article examines different legal outcomes to issues raised by
advances in life support systems technology.

Arthur, Lindsay G. (1981). "In Re Sharon Siebert: Decision Regarding
a Brain-Damaged Adult." Bioethics Quarterly 3 (1): 10-15.

A report of the findings and questions of law raised in the case
of Sharon Siebert, a brain-damaged adult.

Boyajian, Jane A. (1981). "Decision-Making for the Incompetent:
Exemplars and Questions from Siebert." Bioethics Quarterly 3 (1):
3-9.

Boyajian discusses the medical, legal, and ethical issues raised
in the matter of Sharon Siebert, a brain-damaged adult.

Cranford, Ronald E., and Mary S. Schneider (1981). "Siebert
Commentary: Medical-Legal Issues." Bioethics Quarterly 3 (1):
16-20.

A discussion of medical-legal issues in the case of Sharon
Siebert, a brain-damaged adult.

In Re Conroy 464 A. 2d 303 (N.J. Super. A.D. 1983).

This case focuses on the question of whether the state's interest
in the preservation of life outweighs the patient's right of
privacy. If so, withdrawal of nourishment would be an act of
euthanasia, constituting homicide.

Matter of Quinlan (1976). 355 A. 2d 647.

In the case of Karen Ann Quinlan, the patient's right to privacy
outweighed the state's interest in the preservation of life. No
legal liability attached when life support was withdrawn.

25

VI. GENETIC DETERMINANTS OF BEHAVIOR

A. XYY Controversy

Millard v. State (1970). 261 A. 2d. 277.

> In this case, the court debated whether there is a biological determinant of illegal behavior that can be sufficiently demonstrated so as to mitigate the requirement of intent. The court's decision was negative.

People v. Yukl (1975). 372 N.Y.S. 2d 313.

> Further study is required to establish a causal connection between one's genetic complement and a predisposition toward violent criminal conduct. In this court case, it was decided that, until this connection is demonstrated, an individual's genetic makeup is not admissable as part of an insanity defense.

Steinfels, Margaret O., and Carol Levine, eds. (1980). "The XYY Controversy: Researching Violence and Genetics." The Hastings Center Report 10 (4): Suppl. 1-32.

> An edited transcript of a conference that discussed scientific research into genetic determinants of violent behavior. Of special interest is the question of whether an extra Y chromosome reduces liability of violent offenders.

VII. PATENT LAW

Adler, Reid G. (1984). "Biotechnology as an Intellectual Property." Science 224 (4647): 357-363.

> Legal requirements of intellectual property--patents being one type of intellectual property--are discussed as they affect biotechnology.

Beier, F. K., R. S. Crespi, and J. Straus (1985). Biotechnology and Patent Protection: An International Overview. Paris: OECD.

> This report is organized into three sections. Part I presents a general introductory review of patent laws and applications in relation to biotechnology. Part II is based on responses to an OECD questionnaire concerning patent protection in biotechnology by OECD member countries and interested circles. Conclusions and recommendations by a group of experts are presented in Part III. The OECD questionnaire, selected court decisions, international conventions, and an international bibliography are included in the annexes.

26

Bent, Stephen A. (1979). "Living Matter Found to be Patentable: In re Chakrabarty." Connecticut Law Review 11 (2): 311-330.

In this commentary on Chakrabarty, Bent contends that the court did not apply traditional criteria for patentability to the creation of living matter. He examines arguments for and against patentability and discusses conceptual problems raised by both positions.

Cooper, Iver P. (1982). Biotechnology and the Law. New York: Clark Boardman Company, Ltd.

The book covers the patentability of biotechnological invention.

Diamond v. Chakrabarty. 100 S. Ct. 2204 (1980).

Text of the court case in which it was decided that a living, human-made microorganism is patentable under the U.S. Constitution.

Dickson, David (1980). "Inventorship Dispute Stalls DNA Patent Application." Nature 284 (5755): 388.

The patentability of the results of biological research has raised several legal issues. Among those reported by Dickson are the difficulty of determining inventorship and the maintenance of the integrity of the research process.

Holtzman, Irving (1979). "Patenting Certain Forms of Life: A Moral Justification." The Hastings Center Report 9 (3): 9-11.

Commentary on Diamond v. Chakrabarty, in which it was decided that a living, human-made microorganism is patentable.

Plant, David W. (1986). "The Impact of Biotechnology on Patent Law." In Joseph G. Perpich (ed.), Biotechnology In Society: Private Initiatives and Public Oversight. Oxford, Great Britain: Pergamon Press, pp. 125-136.

The impact of biotechnology on the U.S. patent system is the focus of this article. The purpose of the U.S. patent system and its relationship with biotechnology is discussed. The author then examines the issues pertaining to biotechnology and patent law. The impact of the patent system on biotechnology is also considered.

27

Plant, David W., Niels J. Reimers, and Norton D. Zinder, eds. (1982).
  Banbury Report 10: Patenting of Life Forms. Cold Spring Harbor
  Laboratory.

  A collection of papers presented at a conference held in 1981 at
  the Banbury Center of Cold Spring Harbor. These cover scientific
  and legal issues involved in the patenting of life forms.

Saliwanchik, Roman (1982). Legal Protection for Microbiological and
  Genetic Engineering Inventions. Reading, Mass.: Addison-Wesley
  Publishing Co.

  A description of the legal requirements and processing of patent
  applications for biotechnological inventions. Includes
  appendixes and cases.

Schlosser, Stanley D. (1981). "Patenting Biological Inventions." The
  University of Toledo Law Review 12 (4): 925-944.

  This article outlines how various legal systems protect patent
  rights with regard to living matter. It covers the American
  statutory system and foreign systems and treaties.

Sparrow, Calvin N. (1981). "An International Comparative Analysis of
  the Patentability of Recombinant DNA-Derived Organisms." The
  University of Toledo Law Review 12 (4): 945-957.

  Both the U.S. Supreme Court and the European Patent Office have
  decided that DNA-derived microorganisms are patentable. This
  article compares the reasoning behind the decisions.

VIII. TORT LIABILITY

Department of Health, Education, and Welfare: National Institutes of
  Health (1976). "Recombinant DNA Research Guidelines." Federal
  Register 41 (July 7): 27902-27911.

  This draft government document assesses risks and benefits of
  rDNA research and sets guidelines for conduct of such research.
  The chronology leading to the establishment of these guidelines
  is described.

Epstein, Richard A. (1974). "Defenses and Subsequent Pleas in a
  System of Strict Liability." The Journal of Legal Studies 3 (1):
  165-215.

  Epstein examines the legal concept of strict liability.

28

Ford, Brian J. (1974). "Call for Biohazard Legislation." Nature 250 (5465): 364-365.

The author calls for a biohazard law due to problems ranging from lab error or intentional poisoning to leakage from biological warfare experiments.

Friedman, Jane M. (1978). "Health Hazards Associated with Recombinant DNA Technology: Should Congress Impose Liability Without Fault?" Southern California Law Review 51 (6): 1355-1379.

This article discusses tort liability with regard to rDNA technology.

Izenstark, Susan R. (1977). "Genetic Manipulation: Research Regulation and Legal Liability under International Law." California Western International Law Journal 7 (1): 203-227.

The author claims that the nature of genetic research technology creates problems that need to be addressed by an international convention regulating genetic manipulation.

Mahoney v. United States. 220 F. Supp. 823 (1963).

The court's denial of liability in this case was based on the fact that the amount of exposure was within limits fixed by a national committee.

Morris, Clarence (1952). "Hazardous Enterprises and Risk Bearing Capacity." Yale Law Journal 61 (6): 1172-1179.

Morris discusses the doctrine of strict liability, or liability without fault.

Parker v. Employers Mutual Liability Insurance Company of Wisconsin. 440 S.W. 2d 43.

This court decision illustrates that the concept of strict liability fails when measured against the standard of reasonable medical probability. In a dissenting opinion, it is argued that reasonable medical probability is an impossible standard to meet in areas where it is not known what is "medically probable."

Prosser, William L. (1941). Handbook of the Law of Torts. St. Paul, Minn.: West Publishing Company.

Prosser includes in this handbook a statement defining strict liability, together with a summary of the case that created the principle (Rylands v. Fletcher).

29

IX. INTERNATIONAL

Bartley, Robert L., and William P. Kucewicz (1983). "Yellow Rain and the Future of Arms Agreements." Foreign Affairs 61 (4): 805-826.

Evidence of yellow rain in Southeast Asia pointed to a Soviet violation of the Biological Weapons Convention.

Bowman, M. J., and D. J. Harris, comps. and annot. (1984). Multilateral Treaties Index and Current Status. London: Butterworth and Co., pp. 415-416.

This index lists parties to the convention prohibiting environmental modification in warfare.

Hostie, J. F., Charles Warren, and Robert A. E. Greenshields (1941). "Trail Smelter Arbitral Tribunal Decision." In George G. Wilson and George A. Finch (eds.), The American Journal of International Law 35 (4): 684-736.

Responsibility in the event of noncontainment is discussed, presenting the concept that a nation may not use its territory in such a manner as to harm the territory of another state. This argument becomes relevant in the case of escape of rDNA engineered organisms.

Izenstark, Susan R. (1977). "Genetic Manipulation: Research Regulation and Legal Liability Under International Law." California Western International Law Journal 7 (1): 203-227.

See Legal, VIII. Tort Liability, above.

Nanda, Ved P. (1982). "Global Climate Change and International Law." Impact of Science on Society 32 (3): 365-374.

Nanda argues that existing international legal mechanisms are inadequate to deal with the problem of global climatic change.

United Nations (1972). Treaties in International Agreement Series. Convention on International Liability for Damage Caused by Space Objects. TIAS 7762.

This international agreement states that the launching state has absolute liability to pay compensation for damage caused by space objects. Per Raymond A. Zilinskas (1978, see below), this notion of absolute responsibility may apply in the event of noncontainment of rDNA organisms.

_____ (1972). Treaties in International Agreement Series. <u>Convention on the Prohibition of the Development, Production and Stockpiling of Bacteriological (Biological) and Toxin Weapons and on Their Destruction</u>. TIAS 8062.

This is the printed text of the convention.

_____ (1977). Treaties in International Agreement Series. <u>Convention on the Prohibition of Military or any other Hostile Use of Environmental Modification Techniques</u>. TIAS 9614.

The text of the U.S. ratified convention that broadens the Biological and Toxin Weapons Convention to include banning weather modification in warfare.

_____ (1979). "Convention on the Prohibition of Military or any Other Hostile Use of Environmental Modification Techniques." <u>Yearbook of the United Nations 1976</u>. New York: Office of Public Information 30:42-47.

The history of the framing and intent of the convention.

_____ (1981). "Review Conference of the Parties to the Convention on the Prohibition of the Development, Production, and Stockpiling of Bacteriological (Biological) and Toxin Weapons and on Their Destruction." <u>The United Nations Disarmament Yearbook</u>. New York: Department of Political and Security Council Affairs, United Nations Centre for Disarmament 5:258-276.

Text of the conference at which the question of compliance to the convention was discussed.

Weiss, Edith B. (1978). "International Liability for Weather Modification." <u>Climatic Change</u> 1 (3): 267-290.

In this article, Weiss examines the question, "Can a nation-state be held liable to another for its weather modification activities?"

Zilinskas, Raymond A. (1978). "Recombinant DNA Research and the International System." <u>Southern California Law Review</u> 51 (6): 1483-1501.

This article establishes a framework for a discussion of the international implications of biotechnological advances. The author discusses international legal implications of recombinant DNA research for peace and war.

SECTION 3:   ECONOMIC ASPECTS OF BIOTECHNOLOGY

32

SECTION 3: ECONOMIC ASPECTS OF BIOTECHNOLOGY

I. GENERAL

Blank, Robert H. (1981). The Political Implications of Human Genetic
    Technology. Boulder, Colo.: Westview Press.

    Blank presents a summary of current facts and discusses future
    prospects of biotechnology in the social and political context of
    the United States. Alternative value frameworks dealing with the
    issues are presented. Chapter 5, "Assessing Human Genetic
    Programs," is particularly relevant to this section. Issues
    discussed in this chapter include the following: public funding
    for genetic research and technology; cost-benefit analysis in
    public policymaking; cost-benefit applications to genetic
    intervention programs; genetic policymaking; technology
    assessment; and allocation of resources for genetic technology.

Buchanan, Cathy, and Elizabeth W. Prior (1984). "Bureaucrats and
    Babies: Government Regulation of the Supply of Genetic
    Material." The Economic Record 60 (170): 222-230.

    Buchanan and Prior disagree with the utilitarian position that
    ". . .there is no justification for suppliers of genetic material
    receiving the competitive market-clearing price for their product."
    The authors argue this trade practice is not significantly different
    from the sale of other bodily property. They maintain that
    competitive market exchange of genetic material is preferable to
    government regulated distribution ". . .because it does more to
    further individual liberty and because it is more efficient."

Bull, Alan T., Geoffrey Holt, and Malcolm D. Lilly (1982).
    Biotechnology: International Trends and Perspectives. Paris:
    OECD.

    This report examines the field of biotechnology from the
    science, technology, and economic perspective. Furthermore, it
    considers these issues ". . .on a comprehensive international
    scale. . .covering the whole field of biotechnology." These
    issues include the following: potential of contributing sciences
    and technologies to biotechnology; scientific technological, and
    resource constraints on biotechnology; and important issues
    affecting developments in biotechnology. The report also gives
    conclusions and recommendations by a panel of experts,
    appendices, a glossary, and a bibliography.

33

Fudenberg, H. Hugh, ed. (1983). <u>Biomedical</u> <u>Institutions</u>, <u>Biomedical</u>
<u>Funding</u>, <u>and</u> <u>Public</u> <u>Policy</u>. New York: Plenum Press.

This book examines the political and economic implications of
biomedical research. The three areas of focus are the following:
the role of foundations in shaping health policy, and their
differing approaches to biomedical research; economic analyses of
fundamental research within given fields; and past and estimated
future "dollar savings" due to basic biomedical research.

Hardy, Ralph W. F., and David J. Glass (1985). "Our Investment: What
is at Stake?" <u>Issues</u> <u>in</u> <u>Science</u> <u>and</u> <u>Technology</u> 1 (3): 69-82.

This article discounts the concerns of those who argue that the
scientific community has not investigated the potential dangers
that new genetic technologies may pose to human health. The
authors are particularly concerned about regulation of the
biotechnology industry. "If overly stringent regulations are
imposed," they warn, "both a major investment and a rare economic
opportunity will be lost."

Yoxen, Edward (1983). <u>The</u> <u>Gene</u> <u>Business</u>: <u>Who</u> <u>Should</u> <u>Control</u>
<u>Biotechnology</u>? New York: Harper & Row.

In this book, Yoxen addresses the broad social and economic
issues associated with the field of biotechnology. The
importance of the wide range of developments falling under the
term "biotechnology" and the history of molecular biology are
discussed. Furthermore, the science of molecular biology and
genetic engineering are described. Chapters 4, 5, and 6 present
case studies of the role biotechnology plays in the process of
transforming medicine, the agriculture and food industries, and
the production of energy and chemicals, respectively. The
concluding chapter discusses opportunities for a greater public
role in this field. The forward, by Robert M. Young, offers
comments and a complete summary of the book.

II. HUMAN GENETIC INTERVENTION

A.   Genetic Screening

Barden, Howard S., and Raymond Kessel (1984). "The Costs and Benefits
of Screening for Congenital Hypothyroidism in Wisconsin." <u>Social</u>
<u>Biology</u> 31 (3-4): 185-200.

This article presents a cost-benefit analysis of a screening
program for the detection of hypothyroidism in newborns. The
analysis compares the monetary costs of the program with the
projected benefits (avoided costs) that result from early
detection and treatment. The authors conclude that the net
benefits exceed the cost of the program.

34

Barden, Howard S., Raymond Kessel, and Virginia E. Schuett (1984).
     "The Costs and Benefits of Screening for PKU in Wisconsin."
     Social Biology 31 (1-2): 1-17.

     The authors present a cost-benefit analysis of a screening
     program for the detection of PKU in newborns. The analysis
     compares the monetary costs of the program with the projected
     benefits (avoided costs) that result from early detection and
     treatment. The authors conclude that the net benefits exceed the
     cost of the program.

Dagenais, Denyse L., Leon Courville, and Marcel G. Dagenais (1985). "A
     Cost-Benefit Analysis of the Quebec Network of Genetic Medicine."
     Social Science & Medicine 20 (6): 601-607.

     Some serious diseases, including several major genetic disorders,
     may be effectively treated if detection occurs before symptoms
     appear. Systematic population screening can assure such
     detection. The Quebec Network of Genetic Medicine is concerned
     with the sytematic early detection of congenital disorders in
     neonates in Quebec. This article presents a cost-benefit
     analysis of the Network. The general question of the
     socioeconomic profitability of biomedical research is also
     discussed.

United States Congress, Office of Technology Assessment (1983). The
     Role of Genetic Testing in the Prevention of Occupational
     Disease. OTA-BA-194. Washington, D.C.: U.S. Government
     Printing Office.

     This report examines the technology and the social implications
     of genetic testing in the workplace. Chapter 10, "Prospects and
     Problems for the Economic Evaluation of Genetic Testing," is
     particularly relevant to this section. Congressional issues and
     options are also presented.

B. Genetic Therapy

Blank, Robert H. (1981). The Political Implications of Human Genetic
     Technology. Boulder, Colo.: Westview Press.

     See Economics, I. General, above.

Swint, J. Michael (1982). "Antenatal Diagnosis of Genetic Disease:
     Economic Considerations." In Barbara J. McNeil, and Ernest G.
     Cravalho (eds.), Critical Issues in Medical Technology. Boston,
     Mass.: Auburn House Publishing Company.

     Swint provides an economic assessment of current amniocenteses
     use, and its associated benefits. He then describes several
     other prenatal diagnostic technologies and their economic

consequences. The author acknowledges the substantial economic impact of genetic disease and concludes, "As the private insurance carriers bear only a small fraction of the full social cost of genetic disease, it is unlikely that they will be the first to initiate expansion in coverage."

United States Congress, Office of Technology Assessment (1984). _Human Gene Therapy--A Background Paper_. OTA--BP--BA--32. Washington, D.C.: U.S. Government Printing Office.

Issues discussed in this background paper include types and techniques of gene therapy, background information about genetic diseases, issues that may arise from clinical application, social implications of gene therapy, and the federal role in gene therapy. Issues relevant to this section include insurance, federal support of research and payment of therapy, and the fair distribution of benefits and costs. The paper has technical notes, appendices, a glossary, and references.

III. PRENATAL ISSUES

Andreano, Ralph L., and Daniel W. McCollum (1983). "A Benefit-Cost Analysis of Amniocentesis." _Social Biology_ 30 (4): 347-373.

Andreano and McCollum present a cost-benefit analysis of amniocentesis. Both qualitative and quantitative benefits and costs are discussed, and the latter estimated. The analysis ". . .concludes that a program of amniocentesis for all pregnant women beyond age 32 would be warranted." The authors note, however, that "this economic analysis. . .is only one aspect of a policy decision. Moral, legal, and ethical questions must also be considered."

Grobstein, Clifford, Michael Flower, and John Mendeloff (1983). "External Human Fertilization: An Evaluation of Policy." _Science_ 222 (4620): 127-133.

This article presents the current status of public policy in the area of _in vitro_ fertilization. It also provides a discussion of estimated costs, demand, and supply of services concerned with this technology.

36

IV. ISSUES WITHIN THE LIFE CYCLE

A. Organ Transplantation

Annas, George J. (1984). "Life, Liberty, and the Pursuit of Organ Sales." The Hastings Center Report 14 (1): 22-23.

In his examination of organ sales, the author discusses the present law pertaining to this issue, and presents the arguments for and against the sale of organs. He then considers at what level, state or federal, this should be legislated.

_____ (1985). "Regulating the Introduction of Heart and Liver Transplantation." Public Health and the Law 75 (1): 93-95.

This is a summary of a report of the Massachusetts Task Force on Organ Transplantation. Recommendations of the Task Force are presented, including a recommendation of cost-containment mechanisms.

Caplan, Arthur (1983). "Organ Transplants: The Costs of Success." The Hastings Center Report 13 (6): 23-32.

In this article the author considers the following: the emergence and failure of "encouraged voluntarism"; the growing gap between supply and demand; informed and presumed consent; policies to protect living donors; costs of organ transplantation; and the need for evaluation and regulation. The author concludes that existing public policy is inadequate to deal with organ transplantation.

Evans, Roger W. (1986). "The Heart Transplant Dilemma." Issues in Science and Technology 2 (3): 91-101

Evans considers the issue of heart transplant and the arguments against Medicare funding of the procedure. He contends that this procedure is not included in Medicare coverage because of political and economic, not medical, reasons. He concludes that the argument in favor of funding this procedure is as compelling as that for other key treatments. And, if it is decided a new technology is unaffordable, such as heart transplantation, the coverage of other "accepted" therapies with lesser or equal benefits should be reconsidered.

Hellinger, Fred J. (1982). "An Analysis of a Public Program for Heart Transplantation." The Journal of Human Resources 17 (2): 307-313.

This article presents a cost-benefit analysis of a public program for heart transplantation. Cost per transplant, estimated size of a public program to fund heart transplants, and estimated costs and benefits derived from such a program are examined.

37

Reiss, John B., John Burckhardt, and Fred Hellinger (1982). "Costs and Regulation of New Medical Technologies: Heart Transplants as a Case Study." In Barbara J. McNeil, and Ernest G. Cravalho (eds.), Critical Issues in Medical Technology. Boston, Mass.: Auburn House Publishing Company.

The authors examine policy and regulatory issues in context with government payments for new medical procedures and technologies. A case study of heart transplantation provides the research focus. The costs and benefits in relation to social and economic factors are discussed. Potential federal program costs and associated benefits are also considered. The authors then address the ". . .social, ethical, and legal concerns about the regulatory implications of funding heart transplantation."

Rettig, Richard A. (1981). Case Study 1: Formal Analysis, Policy Formulation, and End-Stage Renal Disease. OTA-BP-H-9(1). Washington, D.C.: U.S. Congress, Office of Technology Assessment.

This is the first of a series of case studies provided for the OTA's assessment The Implications of Cost-Effectiveness Analysis of Medical Technology. In this study, the authors analyze ". . .the use of formal analysis in the formulation of Federal Government policy for end-stage renal disease (ESRD)." The case study is presented in the following four sections: background information about ESRD; a review of literature concerned with ESRD, with emphasis on two major analytical studies; the role of those studies in the policy formulation process; and the author's conclusions.

Romeo, Anthony A. (1984). Health Technology Case Study 32: The Hemodialysis Equipment and Disposables Industry. OTA-HCS-32. Washington, D.C.: U.S. Congress, Office of Technology Assessment.

This case study, performed as part of OTA's assessment of Federal Policies and the Medical Devices Industry, presents an analysis of the hemodialysis equipment and supplies industry. The analysis consists of issues including treatment approaches, the market structure and competition, industry performance, and policy issues.

Schwartz, Howard S. (1985). "Bioethical and Legal Considerations in Increasing the Supply of Transplantable Organs: From UAGA to 'Baby Fae'." American Journal of Law & Medicine 10 (4): 397-437.

Organ availability is the major problem associated with organ transplantation. In this article, current organ procurement procedures and technologies, and legislative responses to the scarcity of transplantable organs are among issues examined. Medical costs and benefits, and issues of fairness in the allocation of scarce economic and social resources are also discussed.

38

Schweitzer, S. O., and C. C. Scalzi (1981). Case Study 6: The Cost Effectiveness of Bone Marrow Transplant Therapy and Its Policy Implications. OTA-BP-H-9(6). Washington, D.C.: U.S. Congress, Office of Technology Assessment.

This is one of a series of case studies provided for the OTA's assessment The Implications of Cost-Effectiveness Analysis of Medical Technology. The study presents a description of bone marrow transplant therapy, a cost-effectiveness analysis of the therapy, and policy implications and options. References and appendices are also provided.

Simmons, Roberta G., and Susan Klein Marine (1984). "The Regulation of High Cost Technology Medicine: The Case of Dialysis and Transplantation in the United Kingdom." Journal of Health and Social Behavior 25 (3): 320-334.

The structure and operational policies of the British National Health Services (NHS), with respect to how they affect the allocation of high cost medical technology, are discussed. The research focus of this article is a case study of the technologies of dialysis and kidney transplantation. "The study suggests that a fixed-budget system within a centralized NHS structure in an environment of scarch resources will be associated with shortages, less utilization of expensive technological options in treatment, greater ability of the government to change health priorities and ration care, conflict among interest groups, conflict among and within bureaucratic levels, and more effective cost-control."

Stason, William B., and Benjamin A. Barnes (1985). Health Technology Case Study 35: The Effectiveness and Costs of Continuous Ambulatory Peritoneal Dialysis (CAPD). OTA-HCS-35. Washington, D.C.: U.S. Congress, Office of Technology Assessment.

This case study was prepared in connection with the OTA's project on Medical Technology and Costs of the Medicare Program. In this study, the medical effectiveness of hemodialysis (HD) performed in hospitals and dialysis centers is compared with continuous ambulatory peritoneal dialysis (CAPD) or HD performed at home. The costs of treatment for each of these methods are evaluated, and critical issues requiring further evaluation are identified.

39

B.   The Artificial Heart

Lubeck, Deborah P., and John P. Bunker (1982). Health Technology Case
     Study 9: The Artificial Heart: Cost, Risks, and Benefits.
     OTA-BP-H-9(9). Washington, D.C.: U.S. Congress, Office of
     Technology Assessment.

     This is one of a series of case studies provided for the OTA's
     assessment The Implications of Cost-Effectiveness Analysis of
     Medical Technology. In this study, issues discussed include the
     following: the history of the artificial heart; potential
     recipients; economic aspects; social benefits; and social costs.
     Policy recommendations, appendices, and references are also
     presented.

Preston, Thomas A. (1985). "Who Benefits from the Artificial Heart?"
     The Hastings Center Report 15 (1): 5-7.

     The artificial heart program is certain to have an influence on
     Medicare expenditures, insurance, and the federal health care
     budget. In this article, the author examines the technology's
     claim to therapeutic benefit, its technologic capabilities, and,
     in light of its cost, the need to establish priorities. The
     author argues that the central issue is who will direct the
     policy for this technology.

Shaw, Margery W., ed. (1984). After Barney Clark: Reflections on the
     Utah Artificial Heart Program. Austin, Texas: University of
     Texas Press.

     This book contains the papers presented at a conference organized
     by the University of Utah to examine the social, ethical,
     political, and economic issues surrounding the artificial heart.
     Part four, which discusses the economic, historical, and
     scientific issues, is particularly relevant to this section.
     Appendices and summaries of audience discussions at the
     conference are included.

C.   Blood

Drake, Alvin W., Stan N. Finkelstein, and Harvey M. Sapolsky (1982).
     The American Blood Supply. Cambridge, Mass.: MIT Press.

     The authors examine the American blood supply and provide a
     discussion of issues including the following: alternative blood
     collection ideologies; non-profit organizations; profit-making
     organizations and the plasma sector; cross-national comparisons
     of practices; and identification of important problems and policy
     issues.

40

Eckert, Ross D., and Edward L. Wallace (1985). Securing a Safer Blood Supply: Two Views. Washington, D.C.: American Enterprise Institute for Public Policy Research.

A presentation of two authors studies on the blood supply is the subject of this book. The first, "Blood, Money, and Monopoly," by Eckert provides a discussion of the following: the controversy over cash blood and posttransfusion hepatitis; the American National Red Cross; competition among blood collectors; the AIDS crisis; and new policy approaches. The second study, "The Case for National Blood Policy," by Wallace, discusses the blood services system in the 1970s and 1980s, past and present blood quality, and regionalism, conflicting philosophies, and competition. The author's conclusions differ concerning the volunteer donor system, commercial services, and collection competition.

Hagen, Piet J. (1982). Blood: Gift or Merchandise. New York: Alan R. Liss, Inc.

This book discusses the organization of blood supply, and the various ways that blood services can be organized. Commercial and nonprofit systems are examined, and the pros and cons of each are identified. Appendices are also included.

Klausner, Arthur (1985). "'Adjustment' in the Blood Fraction Market." Bio/Technology 3 (2): 119-125.

This article examines the economics of the blood fractionation industry. Current research, problems, and prospects of this industry are discussed.

Levin, Arthur (1978). "Blood Money." New York 11 (45): 64-68.

The New York Blood Center, a non-profit organization and a large importer of human blood, is the focus of this article. The importation of blood, called the Euroblood program, is discussed. Also, comparisons to commercial blood banks are made.

Titmuss, Richard M. (1971). The Gift Relationship: From Human Blood to Social Policy. New York: Pantheon Books.

This book presents a comparative study of how different societies deal with the problem of procuring human blood for medical purposes. In his study, the author deals with a variety of countries but primarily focuses on the established procurement systems of the U.S. and Great Britain. He examines the voluntary and commercial methods under different systems with relationship to safety, cost per unit, and percentage wasted. The donors within these systems are also discussed, and conclusions concerning their relationship with society are presented.

United States Congress, Office of Technology Assessment (1985). Blood Policy & Technology. OTA-H-260. Washington, D.C.: U.S. Government Printing Office.

Recent developments which created uncertainties in blood banking and transfusion medicine led to the request that the OTA conduct an assessment of blood policy and technologies. This is a report of that assessment. Issues addressed include the following: the blood services complex; costs and availability of blood products; containment of health care costs; alternatives and substitutes for blood products; coordination of blood resources; voluntary vs. commercial sources; and AIDS.

V. AGRICULTURE

Buttel, Frederick H., Martin Kenney, and Jack Kloppenburg, Jr. (1985). "From Green Revolution to Biorevolution: Some Observations on the Changing Technological Bases of Economic Transformation in the Third World." Economic Development and Cultural Change 34 (1): 31-55.

This article examines the emerging "biorevolution" and its implications for Third World agriculture. The differences in the structures of the Green and Biorevolutions are identified. Furthermore, the reinforcement of trends associated with the Green Revolution are discussed. The authors ". . .urge that social research resources be directed to increasing our understanding of the likely impact of biotechnology in developing regions of the globe."

Cook, Kenneth A., and Susan E. Sechler (1985). "Agricultural Policy: Paying for our Past Mistakes." Issues in Science and Technology 2 (1): 97-110.

In this article, the authors examine the transition which agriculture is in, and the major new forces that accompany this transition. One of these forces is technology, including biotechnology. Because of these changes, they maintain old assumptions concerning agriculture no longer hold true. The authors conclude that overproduction is the fundamental problem of U.S. agriculture. Furthermore, they argue, ". . .the cost to government--and ultimately to the taxpayer--will remain high if the United States continues to rely on a fickle export market to revive the farm economy."

Doyle, Jack (1985). Altered Harvest: Agriculture, Genetics, and the Fate of the World's Food Supply. New York: Viking Penguin, Inc.

Doyle examines the uses and implications of biotechnology in the agricultural sector. The ". . .economic race to own the

biological and gentic ingredients of agriculture. . .and who will wield the new genetic ingredients of food power in billion-dollar world markets" are among issues discussed.

_____ (1985). "Biotechnology Research and Agricultural Stability." Issues in Science and Technology 2 (1): 111-148.

This article is concerned with the role of biotechnology in the agricultural sector. The author questions whether the application of agricultural biotechnology will provide broad benefits to agriculture and society, or intensify the weakness of the current high-yield agricultural system. Doyle argues that "biotechnology should be applied in ways that lower costs to farmers, reduce chemical dependency, increase production efficiency, broaden genetic diversity, and enhance biological and economic stability in agriculture here and abroad." He warns, however, that many of these goals are unlikely to be achieved without changes in our agricultural research policy.

Kalter, Robert J. (1985). "The New Biotech Agriculture: Unforeseen Economic Consequences." Issues in Science and Technology 2 (1): 125-133.

The new biotechnology promises unprecedented increases in agricultural productivity. Kalter asserts that, in the long term, this will result in increased agricultural efficiency and improved standards of living. In the short term, however, severe dislocations in some sectors of agriculture may result. Most sensitive to this may be the financially stressed family farms. As new biotech products are introduced into the agricultural sector, new policies may be necessary to minimize the possible economic disruptions.

United States Congress, Office of Technology Assessment (1985). Innovative Biological Technologies for Lesser Developed Countries--Workshop Proceedings. OTA-BP-F-29. Washington, D.C.: U.S. Government Printing Office.

These workshop proceedings examine innovative biological technologies to help lesser developed countries enhance their agricultural productivity. Most of the proceedings are technical in nature. Chapter 2, however, is relevant to this section. This chapter discusses the role of the Agency for International Development (AID) in agricultural development activities, and various constraints under which AID operates. A summary of workshop suggestions is also presented.

United States Congress, Office of Technology Assessment (1985).
   Technology, Public Policy, and the Changing Structure of American
   Agriculture: A Special Report for the 1985 Farm Bill.
   OTA-F-272. Washington, D.C.: U.S. Government Printing Office.

   This report examines the ". . .nature and impacts of emerging
   technologies in combination with public policy specifically as
   they affect the future direction of agriculture." Topics
   discussed in this report include the following: a description of
   the new technologies and their impact on production; economic and
   sociological perspectives of structural change in U.S.
   agriculture; economic and political forces in structural change;
   economic impacts of emerging technologies, and selected farm
   policies for various size crop farms and dairy farms; and
   agricultural research and extension policy. Appendices and
   references are also presented.

Vergopoulos, Kostas (1985). "The End of Agribusiness or the Emergence
   of Biotechnology." International Social Science Journal: Food
   Systems 105 37 (3): 285-299.

   In this article, the author examines the history of agribusiness
   and its significance. This history is outlined in six historical
   theoretical stages: agriculture as an external reserve; the
   social integration of agriculture; integration through
   agribusiness; the organization of the stages of production; the
   economic and food crisis; and the emergence of biotechnology or
   the end of agribusiness. The author then examines possible
   changes in agribusiness, and the effects, resulting from the
   emergence of biotechnology.

VI. INDUSTRIAL

"Biotechnology Becomes a Gold Rush." (1981). The Economist 279
   (7189): 81-86.

   Industries are rapidly seeking the potential profits of
   biotechnology. Although the potential is great, the article
   warns, more research, practical experience, and more investment
   capital than most are contemplating are necessary to realize
   these profits. The article also describes some biotechnological
   discoveries and how industries utilize them.

Glick, J. Leslie (1982). "The Industrial Impact of the Biological
   Revolution." Technology In Society 4:283-293.

   In this article, the author examines the broad impact
   biotechnology holds for many traditional industries. He examines
   the evolution of the current biotechnology industry, which blends
   older technologies with new findings in molecular biology. After

concluding that this merger results in lower costs for raw
materials, labor and capital investment, Glick summarizes the
short- and long-term industrial prospects for the current genetic
engineering industry.

Lappe, Marc (1985). "Biotechnology's Debt to Public Health."
Technology Review 88 (6): 14-15+.

Lappe is concerned that the commercialization of biotechnology is
". . .driven by purely economic motivations," and that most
companies focus ". . .on medical products that promise a high
short-term financial yield. . .designed to alleviate the
disorders of the wealthy, developed countries." He further
argues that this is at the expense of effort to cure widespread
diseases plaguing the third world. He examines several of these
diseases, and the technological capability to combat them. He
provides several policy options to combine humanitarian efforts
with this industry, and concludes that only when this industry
serves these efforts will its enormous potential be fulfilled.

McAuliffe, Sharon, and Kathleen McAuliffe (1981). Life for Sale. New
York: Coward, McCann, and Geoghegan.

In this book, the authors examine the industrialization of
biotechnology. Biotechnology as it relates to the areas of
pharmacology and agriculture, and pioneering firms involved are
among issues discussed. Implications for the developing
countries are also considered. Finally, the safety of this
technology, patents, and possible future directions of
biotechnology are examined.

Newell, Nanette (1986). "International Biotechnology Transfer." In
Joseph G. Perpich (ed.), Biotechnology In Society: Private
Initiatives and Public Oversight. Oxford, Great Britain:
Pergamon Press, pp. 217-219.

The author states, "Transfer of technology is trade dependent
with this country's access to world markets and international
competitiveness." She then describes four methods of
international market penetration: licensing, joint ventures,
subsidiaries, and export. She concludes that open communication
channels, technology transfer, and access to world markets are
necessary to maintain the competitive position of U.S. companies.

Norman, Colin, and Eliot Marshall (1982). "Boom and Bust in
Biotechnology." Science 216 (4550): 1076-1082.

This article provides an account of the financial and legal
troubles of Southern Biotech, a Tampa company that ". . .slipped
from being potentially one of the largest contenders in the race
to commercialize biotechnology, to the brink of bankruptcy." The

implications for the biotechnology industry in general are also considered.

Panem, Sandra (1984). The Interferon Crusade. Washington, D.C.: Brookings Institution.

This book examines the development and promotion of interferon. The author begins with its scientific history and events which led to the industrial race to exploit its market potential. She then discusses the funding of interferon research within the context of the National Cancer Institute and National Institutes of Health cancer research program. Greater financial risks for research support seems to have been borne by private foundations and industry. The author concludes with a brief discussion of policy lessons.

Prentis, Steve (1984). Biotechnology: A New Industrial Revolution. New York: George Braziller, Inc.

Prentis discusses the possible and the actual biotechnological processes and products, particularly in the fields of medicine, agriculture and food production, energy production, and industry. He then examines the potential social, political, and economic effects of these technologies.

Prestowitz, Clyde V., Jr. (1986). "Foreign Targeting in Biotechnology." In Joseph G. Perpich (ed.), Biotechnology in Society: Private Initiatives and Public Oversight. Oxford, Great Britain: Pergamon Press, pp. 197-200.

The author examines the status of the U.S. biotechnology industry and the role of the government for support of this industry. He observes that, except for some research-specific budget allocations, ". . .U.S. government policies provide no direct support for biotechnology development." He then contrasts the U.S. experience with foreign counterparts.

Swanson, Robert A. (1984). "Policies to Stimulate Growth: The View from a New Industry." In National Research Council, The Race for the New Frontier: International Competition in Advanced Technology--Decisions for America.

Swanson examines the biotechnology industry and discusses government policies necessary to stimulate the industry's growth. He concludes, "If government does its part, . . .American business will keep supplying the perspiration and the inspiration that are the common ingredients of innovation."

46

Wade, Nicholas (1980). "Cloning Gold Rush Turns Basic Biology into Big Business." Science 208 (4445): 688-692.

_____ (1980). "Hybridomas: A Potent New Biotechnology." Science 208 (4445): 692-693.

_____ (1980). "Three New Entrants in Gene Splicing Derby." Science 208 (4445): 690.

The above three short articles examine the commercialization of molecular biology and its present stage in industry. The technology of monoclonal antibodies is also discussed.

Yoxen, Edward (1983). The Gene Business: Who Should Control Biotechnology? New York: Harper & Row.

See Economics, I. General, above.

VII. INTERNATIONAL

Bull, Alan T., Geoffrey Holt, and Malcolm D. Lilly (1982). Biotechnology: International Trends and Perspectives. Paris: OECD.

See Economics, I. General, above.

Lakoff, Sanford (1984). "Biotechnology and the Developing Countries." Politics and the Life Sciences 2 (2): 151-187.

This article is concerned with the possible socioeconomic benefits of biotechnology for the developing countries. The author provides a description of several biotechnology techniques, and discusses the economic problems of the developing countries. He then explores the possible economic, social, and political consequences associated with the use of biotechnology in the developing world.

Newell, Nanette (1986). "International Biotechnology Transfer." In Joseph G. Perpich (ed.), Biotechnology In Society: Private Initiatives and Public Oversight. Oxford, Great Britain: Pergamon Press, pp. 217-219.

See Economics, VI. Industrial, above.

Prestowitz, Clyde V., Jr. (1986). "Foreign Targeting in Biotechnology." In Joseph G. Perpich (ed.), Biotechnology in Society: Private Initiatives and Public Oversight. Oxford, Great Britain: Pergamon Press, pp. 197-200.

See Economics, VI. Industrial, above.

United States Congress, Office of Technology Assessment (1984).
   Commercial Biotechnology: An International Analysis.
   OTA-BA-218. Washington, D.C.: U.S. Government Printing Office.

   Topics covered include the industrial use of rDNA, cell fusion
   and novel bioprocessing techniques for agriculture,
   pharmaceuticals, energy, and bioelectronics. The competitive
   position of the United States in this field with that of five
   other states is assessed. Separate chapters discuss the relative
   importance of ten factors that determine a state's competitive
   position. Policy options are identified with respect to these
   factors. Chapters 4, 12, 13, and 19 are of special interest,
   which discuss the following: firms commercializing biotechnology;
   financing and tax incentives for firms; government funding of
   basic and applied research; and international technology
   transfer, investment, and trade. The book has a summary, a
   description of technologies, and excellent appendices.

United States Congress, Office of Technology Assessment (1985).
   Innovative Biological Technologies for Lesser Developed
   Countries--Workshop Proceedings. OTA-BP-F-29. Washington, D.C.:
   U.S. Government Printing Office.

   See Economics, V. Agriculture, above.

_____ (1985). Status of Biomedical Research and Related Technology
   for Tropical Diseases. OTA-H-258. Washington, D.C.: U.S.
   Government Printing Office.

   This report examines the status of biomedical research and
   technologies for controlling tropical diseases. The political,
   scientific, and economic objectives of U.S. involvement in this
   area are discussed. Funding sources, and types of research
   funded are also described. Also, findings and options are
   presented. The report includes two case studies, a glossary, and
   references.

Ventura, Arnoldo K. (1982). "Biotechnologies and Their Implications
   for Third World Development." Technology In Society 4:109-129.

   Ventura maintains that the capacity to develop useful microbes
   through biotechnology offers both the developed and developing
   countries new socioeconomic opportunities. He argues, however,
   that their existence as a commerical substance will exert a heavy
   burden on the developing countries because they lack the
   necessary skills to exploit their use.

48

Yuthavong, Yongyuth, Chatri Sripaipan, Krissanapong Kirtikara, Atthakorn Glankwamdee, and Kanchana Trakulku (1985). "Key Problems in Science and Technology in Thailand." Science 227 (4690): 1007-1011.

The authors describe a dichotomy that exists in Thailand. Thailand's economy is now sophisticated enough to require an expanded role for science and technology. However, it has not yet developed a policy for expanding its scientific facilities and personnel. The authors argue Thailand must strengthen the infrastructure of its science and technology in the following ways: encourage students to pursue science related fields; develop a plan for obtaining more technology transfer; and expand R & D efforts. The authors suggest that Thailand focus on the development of bioscience and biotechnology, as well as several other areas.

SECTION 4: INTERNATIONAL BIOTECHNOLOGY

SECTION 4: INTERNATIONAL BIOTECHNOLOGY

I. BIOLOGICAL TERRORISM

Alexander, Yonah (1981). "Super-Terrorism." In Yonah Alexander and
    John M. Gleason (eds.), Behavioral and Quantitative Perspectives
    on Terrorism. New York: Pergamon Press, pp. 343-361.

    Alexander discusses the possible terrorist use of chemical,
    biological, and nuclear weapons.

Kupperman, Robert H., and Darrell M. Trent (1979). Terrorism: Threat,
    Reality, Response. Stanford, Calif.: Hoover Institution Press.

    The introduction and first three chapters of this volume are
    relevant to biological terrorism. Included in the discussion
    are possible terrorist use of biological agents to achieve aims,
    and the management, control, and safeguards against this
    possibility.

Livingstone, Neil C. (1982). The War Against Terrorism. Lexington,
    Mass.: Lexington Books.

    The book is a survey of various concerns about international
    terrorism. Of special interest is chapter six, "Terrorist
    Weapons: Today and Tomorrow," which includes a discussion of
    terrorist arsenals, and possible terrorist use of
    chemical/biological weapons.

Mullen, Robert K. (1978). "Mass Destruction and Terrorism." Journal
    of International Affairs 32 (1): 63-89.

    This volume discusses development of biological, chemical, and
    nuclear weapons in the context of the possibility that terrorists
    will acquire the means to inflict mass destruction.

Oots, Kent L., and Thomas C. Wiegele (1985). "Terrorist and Victim:
    Psychiatric and Physiological Approaches From a Social Science
    Perspective." Terrorism 8 (1): 1-32.

    The authors present a biosocially interactive model of terrorist
    causation at the individual level of analysis. The article gives
    reasons why violence that may be rooted in the physiology of the
    terrorist has a contagion effect, and it discusses policy and
    methodological issues raised by the psychophysiological approach.

51

Silverstein, Martin E. (1979). "The Medical Survival of Victims of Terrorism." In Robert H. Kupperman and Darrell M. Trent (eds.), Terrorism: Threat, Reality, Response. Stanford, Calif.: Hoover Institution Press, pp. 349-392.

The author discusses the medical response to terrorist use of biological or chemical agents. Beyond the medical, the article raises policy considerations.

Zimmerman, Burke K. (1984). "Directing the Course of Biotechnology: Biological Weapons." Politics and the Life Sciences 2 (2): 188-191.

It is in the nature of biotechnical research that properties that are effective for beneficial purposes are also effective for military purposes. Zimmerman argues that it may soon be possible to create an ethnic or population specific weapon. In addition, he asserts that it is virtually impossible to tell a biological weapons research facility from a "peaceful" molecular biology laboratory, thereby complicating inspection procedures.

II. BIOLOGICAL WARFARE

Bartley, Robert L., and William P. Kucewicz (1983). "Yellow Rain and the Future of Arms Agreements." Foreign Affairs 61 (4): 805-826.

This article reviews the use of toxins which are poisons produced by biological processes that are not themselves living organisms. While the Biological and Toxin Weapons Convention prohibits development and stockpiling of biological or toxin weapons, technology may blur the line between biological and chemical by allowing toxins to be made synthetically. Whether these are chemical or biological, Soviet use of "yellow rain" is biochemical warfare that violates the 1975 Convention or the 1925 Geneva Protocol.

Kaplan, Martin M. (1983). "Another View." Bulletin of the Atomic Scientists 39 (9): 27.

A rejoinder to a previously published article in which the author states that rDNA technology is not important for biological warfare.

Stockholm International Peace Research Institute (1971). "Instances and Allegations of CBW, 1914-1970." In SIPRI, The Problem of Chemical and Biological Warfare. Vol. 1, The Rise of CB Weapons. New York: Humanities Press, pp. 125-230.

In the early 1970s the SIPRI began producing a series of volumes that are a comprehensive survey of all aspects of chemical and

biological warfare. These are directed at scholars and policy makers. This portion is a history of all instances known between 1914 and 1970 in which CB weapons have been used or their use has been alleged.

_____ (1971). "Popular Attitudes towards CBW, 1919-1939." The Problem of Chemical and Biological Warfare. Vol. 1, The Rise of CB Weapons. New York: Humanities Press, pp. 231-267.

Chemical and biological weapons were not used in World War II. This is an examination of popular attitudes during the time preceding that war.

_____ (1977). "Chemical and Biological Weapons." In SIPRI, Weapons of Mass Destruction and the Environment. London: Taylor and Francis Ltd., pp. 31-48.

An examination of environmental aspects of biological, chemical, and nuclear arms control that is directed at policy makers.

Tucker, Jonathan B. (1984). "Gene Wars." Foreign Policy 57 (Winter 1984-85): 58-79.

The biotechnology of gene splicing to alter the practical liabilities of biological warfare is discussed. Although contrary to the Biological and Toxin Weapons Convention, various scenarios become possible with the result that the logic of deterrence becomes attractive.

United Nations (1972). Treaties in International Agreement Series. Convention on the Prohibition of the Development, Production, and Stockpiling of Bacteriological (Biological) and Toxin Weapons and on Their Destruction. TIAS 8062.

This is the text of the convention.

_____ (1980). Report of the Preparatory Committee for the Review Conference of the Parties to the Convention on the Prohibition of the Development, Production, and Stockpiling of Bacteriological (Biological) and Toxin Weapons and on Their Destruction. BWC/Conf. I/5 (80-03360). Geneva, March 3-21.

A background paper on new scientific and technological developments that are relevant to compliance with the convention.

53

_____ (1981). "Review Conference of the Parties to the Convention on the Prohibition of the Development, Production and Stockpiling of Bacteriological (Biological) and Toxin Weapons and on Their Destruction." The United Nations Disarmament Yearbook. Vol. 5. New York: Department of Political and Security Council Affairs, United Nations Centre for Disarmament, pp. 258-276.

This is the text of the conference at which compliance to the Convention was discussed.

Wright, Susan, and Robert L. Sinsheimer (1983). "Recombinant DNA and Biological Warfare." Bulletin of the Atomic Scientists 39 (9): 20-26.

While deliberate construction of harmful biological agents by means of rDNA technology is prohibited, the nature of peaceful or defensive research is to blur the line in a way that may breach the intent of the Convention. These issues are discussed by Wright and Sinsheimer.

Zilinskas, Raymond A. (1983). "New Biotechnology: Potential Problems, Likely Promises." With commentaries and author's response. Politics and the Life Sciences 2 (1): 42-75.

Concluding that biotechnology may overcome the practical limitation of biological conflict, Zilinskas here argues that verification of treaty compliance is needed. In addition, he discusses biopolicy, international organizations, and technology transfer. Beyond this, the author discusses the role of the social scientist in bridging the gap between science and politics through examining the policy implications of new biotechnology. The use of international organizations in technology transfer and the problem of biological warfare are also considered. Commentaries expand on arguments made.

_____ (1983). "Anthrax in Sverdlovsk?" Bulletin of the Atomic Scientists 39 (6): 24-27.

Zilinskas discusses outbreak of anthrax in Sverdlovsk, which raised questions about Soviet compliance with the Biological and Toxin Weapons Convention. He concludes that agreement on verification is needed.

Zimmerman, Burke K. (1984). "Directing the Course of Biotechnology: Biological Weapons." Politics and the Life Sciences 2 (2): 188-191.

See International, I. Biological Terrorism, above.

54

III. CLIMATIC MANIPULATION

Bowman, M. J., and D. J. Harris, comps. and annot. (1984).
     Multilateral Treaties Index and Current Status. London:
     Butterworth and Co., pp. 415-416.

     This reference book lists parties to the convention prohibiting
     environmental modification in warfare.

Nanda, Ved P. (1982). "Global Climate Change and International Law."
     Impact of Science on Society 32 (3): 365-374.

     Nanda argues that existing international legal mechanisms are
     inadequate to deal with the problem of global climatic change.

Panchev, S. (1982). "Intentional Modification of the Atmosphere."
     Impact of Science on Society 32 (3): 355-363.

     In this article, Panchev discusses large-scale experimental
     modification of the atmosphere. According to the author, the
     result of this modification could be catastrophic and
     irreversible. In discussing regional experiments in the Soviet
     Union, the author claims 80 percent effectiveness for operational
     protection of crops from hailstorms. Attempts to produce rain
     for crops have been less successful, although the point is made
     that this capability could lead to international conflict, since
     clouds may produce rain in one region at the expense of another.

Rogers, Kenneth A. (1982). The International Implications of Weather
     and Climate Modification. Ph.D. dissertation, The American
     University. Ann Arbor, Mich.: University Microfilms
     International. DA 8226711.

     This Ph.D. dissertation presents a global commons issue in which
     institutional adjustments lag behind technological innovation and
     environmental degradation. Climate modification, whether
     intentional or inadvertent, requires global management by
     international authority before it becomes a volatile political
     issue.

Schneider, Stephen H., and Lynne E. Mesirow (1976). The Genesis
     Strategy. New York: Plenum Press.

     Schneider and Mesirow present a survey of concerns about climate,
     technology, and human survival, discussing climate-related issues
     from a world perspective. Among other topics, the book discusses
     albedo, greenhouse effect, human impact, deliberate modification,
     public policy, and military applications, offering a strategy
     that includes terrorist abatement and other remedies.

55

Stockholm International Peace Research Institute (1977). "Geophysical and Environmental Weapons." In SIPRI, Weapons of Mass Destruction and the Environment. London: Taylor and Francis Ltd., pp. 49-63.

This article discusses geophysical warfare, that is, the use of weather as a weapon in order to modify the environment to the advantage of one side and to the disadvantage of the other. The techniques used--fire, flood, and rainmaking--lend themselves to covert application and result in severe ecological problems, both immediate and delayed.

Tickell, Crispin (1977). Climatic Change and World Affairs. Harvard Studies in International Affairs, Number 37. Harvard University: Center for International Affairs.

According to Tickell, natural and unnatural variation in climate is an international problem. The book discusses the science and politics of climatic change and recommends international coordination to prevent abuse.

United Nations (1977). Treaties in International Agreement Series. Convention on the Prohibition of Military or Any Other Hostile Use of Environmental Modification Techniques. TIAS 9614.

The text of the U.S. ratified convention that broadens the Biological and Toxin Weapons Convention to include banning weather modification in warfare.

_____ (1979). "Convention on the Prohibition of Military or Any Other Hostile Use of Environmental Modification Techniques." Yearbook of the United Nations 1976. Vol. 30. New York: Office of Public Information, pp. 42-47.

The history of the framing and intent of the convention is presented.

Weiss, Edith B. (1978). "International Liability for Weather Modification." Climatic Change 1 (3): 267-290.

In this article, Weiss examines the question, "Can a nation-state be held liable to another for its weather modification activities?"

IV. INSTITUTIONAL ADJUSTMENTS

Bowman, M. J., and D. J. Harris, comps. and annot. (1984). Multilateral Treaties Index and Current Status. London: Butterworth and Co. pp. 415-416.

See International, III. Climatic Manipulation, above.

56

Hostie, J. F., Charles Warren, and Robert A. E. Greenshields (1941). "Trail Smelter Arbitral Tribunal Decision." In George G. Wilson and George A. Finch (eds.), The American Journal of International Law 35 (4): 684-736.

Responsibility in the event of noncontainment is discussed, presenting the concept that a nation may not use its territory in such a manner as to harm the territory of another state. This argument becomes relevant in the case of escape of rDNA engineered organisms.

United Nations (1972). Treaties in International Agreement Series. Convention on International Liability for Damage Caused by Space Objects. TIAS 7762.

This international agreement states that the launching state has absolute liability to pay compensation for damage caused by space objects. This notion of absolute responsibility may apply in the event of noncontainment of rDNA organisms.

_____ (1972). Treaties in International Agreement Series. Convention on the Prohibition of the Development, Production, and Stockpiling of Bacteriological (Biological) and Toxin Weapons and on Their Destruction. TIAS 8062.

See International, II. Biological Warfare, above.

_____ (1977). Treaties in International Agreement Series. Convention on the Prohibition of Military or Any Other Hostile Use of Environmental Modification Techniques. TIAS 9614.

See International, III. Climatic Manipulation, above.

_____ (1979). "Convention on the Prohibition of Military or Any Other Hostile Use of Environmental Modification Techniques." Yearbook of the United Nations 1976. Vol. 30. New York: Office of Public Information, pp. 42-47.

See International, III. Climatic Manipulation, above.

_____ (1980). Report of the Preparatory Committee for the Review Conference of the Parties to the Convention on the Prohibition of the Development, Production, and Stockpiling of Bacteriological (Biological) and Toxin Weapons and on Their Destruction. BWC/Conf. I/5 (80-03360). Geneva, March 3-21.

See International, II. Biological Warfare, above.

57

_____ (1981). "Review Conference of the Parties to the Convention on the Prohibition of the Development, Production, and Stockpiling of Bacteriological (Biological) and Toxin Weapons and on Their Destruction." The United Nations Disarmament Yearbook. Vol. 5. New York: Department of Political and Security Council Affairs, United Nations Centre for Disarmament, pp. 258-276.

See International, II. Biological Warfare, above.

Weiss, Edith B. (1979). "International Legal Aspects of Recombinant DNA Research." In Joan Morgan and W. J. Whelan (eds.), Recombinant DNA and Genetic Experimentation. New York: Pergamon Press Ltd., pp. 245-259.

In this article, the author discusses international responsibilities to create a legal regime that will spread benefits, minimize risks, and settle disputes arising out of rDNA research.

Zilinskas, Raymond A. (1978). "Recombinant DNA Research and the International System." Southern California Law Review 51 (6): 1483-1501.

This article establishes a framework for a discussion of the international implications of biotechnological advances. The author discusses international legal implications of recombinant DNA research for peace and war.

V. ORGANIZATIONS

Haas, Ernst B., Mary P. Williams, and Don Babai (1977). Scientists and World Order. Berkeley: University of California Press.

The authors discuss the uses of technical knowledge in international organizations and evaluate the implications of scientists' involvement in international agencies.

International Centre for Genetic Engineering and Biotechnology (1984). Conclusions and Decisions. ICGEB/Prep. Comm./4/8 (V. 84-91440). Vienna, Austria, September 17-19.

Decisions made by the Preparatory Committee on the Establishment of the International Centre for Genetic Engineering and Biotechnology.

58

UNIDO Secretariat (1984). Information Note on the Activities of UNIDO's Technology Group in the Field of Genetic Engineering and Biotechnology. Vienna, Austria, September 17-19.

A discussion of UNIDO activities in the field of genetic engineering and biotechnology that was prepared to aid in the establishment of an international center for those fields.

_____ (1984). Review of the Progress in Relation to the Work Programme of the Preparatory Committee on the ICGEB. ICGEB/Prep. Comm./5/4 (V. 84-92906). Trieste, Italy, December 3-5.

An evaluation of the progress made on the Conclusions and Decisions adopted earlier by the International Centre for Genetic Engineering and Biotechnology (ICGEB). This includes scientific advisors, project leader, statutes, recruitment, and financial contributions.

_____ (1984). Progress Report on the Affiliated Centres of the International Centre for Genetic Engineering and Biotechnology. ICGEB/Prep. Comm./5/6 (V. 84-93061). Trieste, Italy, December 3-5.

A listing of the core of information needed from candidate institutions to enable decision making regarding their possible affiliation with the ICGEB and a description of proposals from nominating countries.

_____ (1984). Draft Work Programme for the First Five Years of Operation of the International Centre for Genetic Engineering and Biotechnology. ICGEB/Prep. Comm./5/2 (V. 84-93022). Trieste, Italy, December 3-5.

Following the reclassification of the ICGEB as a permanent rather than a provisional U.N. agency, this document was prepared to present proposals for the work program and a budget for its implementation. The Trieste component focuses on applied research in industrial biotechnology including the conversion of biomass and protein engineering. The New Delhi component focuses research and development on agriculture, livestock productivity, and human health.

United Nations Environmental Program (1981). "Conservation/Living and Genetic Resources." United Nations Environmental Program Annual Review 1981. Nairobi, Kenya: UNEP, pp. 57-63.

A report on the program's conservation of genetic resources. The program is needed because domestication has the effect of depriving agricultural and animal populations of genetic interaction with their wild relatives. In addition, "domestication" of microorganisms makes possible the controlled

recycling of organic matter, elimination of pollutants, and
control of pests and pathogens.

Weiss, Charles, and Nicolas Jequier, eds. (1984). Technology,
Finance, and Development. Lexington, Mass.: Lexington Books.

The role of the World Bank in technology transfer to the
developing world is discussed in this volume. Part I examines
how technology is mobilized at the sector level for agriculture
and industry. Part II focuses on appropriate technology for the
poor. Part III discusses ways in which technology is transferred,
resulting in the buildup of local capabilities. Part IV
describes ways in which the Bank has contributed, financially and
organizationally, to building international research networks.
The editors do not focus exclusively on biotechnology, but
discuss the transfer of technology of all kinds.

Zilinskas, Raymond A. (1983). "New Biotechnology: Potential Problems,
Likely Promises." With commentaries and author's response.
Politics and the Life Sciences 2 (1): 42-75.

See International, II. Biological Warfare, above.

VI.   TECHNOLOGY TRANSFER

Agnew, John A. (1982). "Technology Transfer and Theories of
Development: Conceptual Issues in the South Asian Context."
Journal of Asian and African Studies 17 (1-2): 16-31.

The author presents a survey of different theories of development
that result in different perspectives on technology transfer.

Almond, Gabriel, and Sidney Verba (1965). Civic Culture. Boston:
Little, Brown and Co.

In chapter one, the authors provide a framework for the concept
that emerging nations need to develop congruence between
political culture and political institutions. The same argument
holds that technological institutions may require an appropriate
culture that is characterized by a moderate mix of the old and
the new.

Birkner, John H. (1986). "Biotechnology Transfer--National Security
Implications." In Joseph G. Perpich (ed.), Biotechnology in
Society: Private Initiatives and Public Oversight. Oxford,
Great Britain: Pergamon Press, pp. 201-207.

The basis for adding biotechnology to the Militarily Critical
Technologies List is the focus of this article. Critical
technologies are discussed and some perceptions of those in the

60

Department of Defense are considered. Finally, future concerns and some views of cooperation are presented.

de Bettignies, H. C. (1978). "The Management of Technology Transfer: Can It Be Learned?" Impact of Science on Society 28 (4): 321-327.

According to de Bettignies, technology transfer is disruptive because it brings a variety of interdependent economic, social, and political changes. A desire on the part of developing countries to learn how to manage this transfer leads to a North-South dialogue that boils the issue down to one of acquiring the "best" technology on conditions that are the least "unfair."

Desa, V. G. (1978). "Research Coordination and Funding Agencies in Developing Countries." Impact of Science on Society 28 (2): 105-116.

To aid in the application of science and technology as a strategy for development, Desa offers an institutional framework for planning and implementing appropriate technologies.

Forje, John W. (1978). "Poor Nations Need to Bargain for a Better Deal in Development." Impact of Science on Society 28 (2): 193-197.

In this article, written from the point of view of developing countries, the author postulates a development wheel of "economic imperialism." He concludes with a call for a negotiating strategy that will get the right kinds of technology from developed nations.

Godet, Michel (1983). "Crisis and Opportunity: From Technological to Social Change." Futures 15 (4): 251-263.

According to Godet, decisions made in respect to new technologies affect the international economy. The author feels that, in order to avoid being in perpetual crisis, we must recognize that technology creates new "rules of the game." Technological pluralism is recommended.

Malik, Yogendra K., and Surinder M. Bhardwaj (1982). "Politics, Technology, and Bureaucracies: An Overview." Journal of Asian and African Studies 17 (1-2): 1-15.

The entire issue is about technology transfer for development. The authors define terms and frameworks for a discussion of technology transfer for development and make a connection between socio-cultural background and technological choice.

61

Malik, Yogendra K. (1982). "Attitudinal and Political Implications of Diffusion of Technology: The Case of North Indian Youth." Journal of Asian and African Studies 17 (1-2): 45-73.

In this study, a high degree of technological exposure is positively correlated with a moderate political orientation and a preference for consumer goods.

Newell, Nanette (1986). "International Biotechnology Transfer." In Joseph G. Perpich (ed.), Biotechnology in Society: Private Initiatives and Public Oversight. Oxford, Great Britain: Pergamon Press, pp. 217-219.

The author states, "Transfer of technology is trade dependent with this country's access to world markets and international competitiveness." She then describes four methods of international market penetration: licensing, joint ventures, subsidiaries, and export. She concludes that open communication channels, technology transfer, and access to world markets are necessary to maintain the competitive position of U.S. companies.

Perez, Carlota (1983). "Structural Change and Assimilation of New Technologies in the Economic and Social Systems." Futures 15 (5): 357-375.

Perez argues that current social and institutional structures cannot assimilate emerging technologies. Dynamic change produces disruption which demands complementary innovation in the social and institutional spheres. This disruption causes economic upturns and downturns.

Swaminathan, M. S. (1982). "Biotechnology Research and Third World Agriculture." Science 218 (4576): 967-972.

An assessment of potential applications of biotechnology as they affect agriculture in developing countries. This includes the use of genetic engineering to incorporate nitrogen-fixing genes into rice. Agencies for technology transfer are also discussed.

United States Congress, Office of Technology Assessment (1984). Commercial Biotechnology: An International Analysis OTA-BA-218. Washington, D.C.: U.S. Government Printing Office.

Topics covered include the industrial use of rDNA, cell fusion and novel bioprocessing techniques for agriculture, pharmaceuticals, energy, and bioelectronics. The competitive position of the United States in this field with that of five other states is assessed. Separate chapters discuss the relative importance of 10 factors that determine a state's competitive position. Chapter 19 is of special interest: International Technology Transfer, Investment, and Trade. The book has a summary, a description of technologies, and appendices.

62

Vajpeyi, Dhirendra (1982). "Modernity and Industrial Culture of Indian Elites." Journal of Asian and African Studies 17 (1-2): 74-97.

Vajpey explores the relationship between elite belief structure and consequences for industrialization. The focus is on policy culture aspects from the cognitive, affective, and evaluative viewpoints.

Weiss, Charles, and Nicolas Jequier, eds. (1984). Technology, Finance, and Development. Lexington, Mass.: Lexington Books.

See International, V. Organizations, above.

Yapa, Lakshman S. (1982). "Innovation Bias, Appropriate Technology, and Basic Goods." Journal of Asian and African Studies 17 (1-2): 32-44.

Innovation bias may be technological, social, or ecological, according to Yapa. The goal of development is to select an appropriate technology that provides basic goods for masses of people in a way that avoids adverse ecopolitical consequences. This is a conceptual paper within an ecopolitical framework.

Zevin, Leon Z. (1978). "An Integrated Approach to Technology Transfer: Soviet Cooperation with Developing Countries." Impact of Science on Society 28 (2): 183-192.

The fact that developed countries have different social systems affects the character of technology transfer. This article by a Soviet scientist offers bilateral and multilateral examples of an integrated approach to technology transfer.

Zilinskas, Raymond A. (1983). "New Biotechnology: Potential Problems, Likely Promises." With commentaries and author's response. Politics and the Life Sciences 2 (1): 42-75.

See International, II. Biological Warfare, above.

VII. AGRICULTURE AND ENVIRONMENT

Ayanaba, A. (1982). "Bacteria and the Nitrogen Economy." Impact of Science on Society 32 (2): 179-187.

In the developing world, according to Ayanaba, it will be more economical to use nitrogen-fixing bacteria for crop production than to rely on present use of chemically-fixed nitrogen by means of chemical fertilizer. An introduced bacterium will remain in the soil year after year, thereby removing the need for yearly fertilization. Both "old" and "new" biotechnological research is directed to this end.

63

Barton, Kenneth A., and Winston J. Brill (1983). "Prospects in Plant
    Genetic Engineering." Science 219 (4585): 671-675.

    Current agricultural technology causes narrowing of the genetic
    base and accumulation of toxic residue from fertilizer,
    pesticides, and herbicides, according to the authors. Although
    not yet operational, recombinant DNA technology offers the
    potential for a range of genetic modifications that will solve
    these problems.

Chapin, Georganne, and Robert Wasserstrom (1981). "Agricultural
    Production and Malaria Resurgence in Central America and India."
    Nature 293 (5829): 181-185.

    Chapin and Wasserstrom discuss the biosocial effects of old
    technology (chemical pesticides) in comparison with biosocial
    effects of "new" biotechnology (microbial insecticides). See
    also Miller, Lingg, and Bulla (1983), this section.

Dahlberg, Kenneth A. (1979). Beyond the Green Revolution. New York:
    Plenum Press.

    Negative consequences of the green revolution lead to Dahlberg's
    assessment of the history, current uses, and future prospects of
    Western agricultural technology in underdeveloped areas.
    Contextual analysis is used to sort out problems of agricultural
    development within a conceptual framework that takes account of
    evolution and ecology. Although the author does not discuss
    biotechnology specifically, the book provides a basis of
    evaluating whether or how to introduce biotechnical solutions for
    the problems of agricultural development. Policy, risk
    assessment, and appropriate technologies are considered.

Da Silva, Edgar J., Reuben Olembo, and Anton Burgers (1978).
    "Integrated Microbial Technology for Developing Countries:
    Springboard for Economic Progress." Impact of Science on Society
    28 (2): 159-182.

    The authors summarize applications of microbial biotechnology as
    they affect the food, fuel, and fertilizer needs of developing
    countries.

Da Silva, E. J., W. Shearer, and B. Chatel (1980). "Renewable
    Bio-Solar and Microbial Systems in 'Eco-Rural' Development."
    Impact of Science on Society 30 (3): 225-234.

    Research programs to derive fuel from renewable biomass offer a
    promise of solar-based photobiological and microbial energy
    producing systems that can be generated by decentralized
    technology at low cost, according to the authors. This may
    involve the creation of energy crops or the use of crop residue.

64

El Nawawy, Amin S. (1982). "The Promise of Microbial Technology." Impact of Science on Society 32 (2): 157-167.

Both "old" and "new" biotechnologies are able to put bacteria to work to solve the world food and energy problems. El Nawawy discusses nitrogen fixation for crop fertilization; microbial protein for livestock feed; microbial synthesis of antibiotics; and biogas from dung or fermentation of agricultural biomass. This article is a good introduction to the topic of biotechnology and agriculture.

Hill, H. Monte (1984). "Biotechnology: The Key to Third World Development." Politics and the Life Sciences 3 (1): 84-86.

In this commentary, the author makes the point that biotechnology is a necessary but not sufficient condition for development. To be sufficient, Third World nations must bargain intelligently before granting access to their markets and raw materials. Environmental change brought about by biotechnology needs to be accompanied by changes in culture.

Lakoff, Sanford (1984). "Biotechnology and the Developing Countries." With commentaries and author's response. Politics and the Life Sciences 2 (2): 151-187.

Biotechnology creates the prospect of improved plant and animal food production in less developed countries. Economic development may be caused by improved disease control, microbial herbicides, hybrids from species that are sexually incompatible, nitrogen fixation in the absence of fertilizer, and production of fuel from biomass. The author evaluates social costs that result in political consequences for North-South relations, as multinationals become distributors of the technology. A partial solution may be the establishment of multilateral agencies for technology transfer. Commentaries and author's response elaborate on points made in the article.

Miller, Lois K., A. J. Lingg, and Lee A. Bulla, Jr. (1983). "Bacterial, Viral, and Fungal Insecticides." Science 219 (4585): 715-721.

Biotechnology may create genetically engineered bacteria and viruses that make "biological warfare" on insect pests that currently destroy agricultural crops. Use of microbial insecticides may also eliminate toxic pollution created by chemical pesticides. In this article the authors do not consider benefits to Third World agriculture, but that connection is easily made.

65

Morita-Lou, Hiroko (1985). <u>Science</u> <u>and</u> <u>Technology</u> <u>Indicators</u> <u>for</u> <u>Development</u>. Boulder, Colo.: Westview Press.

    The contributors to this book contend that the current indicators used in the measurement of the impact of science and technology in developing countries ". . .have been formulated based on conditions and assumptions that are primarily relevant to developed countries." When these indicators are applied to developing countries, the conclusions drawn are often inaccurate. These issues, including assessment of qualitative indicators, in context with developing countries, is the focus of this book. Case studies of current programs in Brazil and China are also included.

Moulik, Tushar K. (1981). "Biogas: The Indian Experience." <u>The</u> <u>UNESCO</u> <u>Courier</u> 34 (7): 33-34.

    With an input of dung, a small biogas digester can produce two to three meters of gas per day. According to Moulik, Indian cultural values and economic constraints prevent full success of the program.

Silva, Benedicto (1981). "Brazil's Green Gasoline." <u>The</u> <u>UNESCO</u> <u>Courier</u> 34 (7): 32.

    Brazil may become the first country in the world to become self-sufficient in renewable fuel. This article is in an issue of the <u>UNESCO</u> <u>Courier</u> devoted to the world energy problem, and it offers one biotechnical solution.

Sineriz, Faustino (1982). "Microbial Fuel Production." <u>Impact</u> <u>of</u> <u>Science</u> <u>on</u> <u>Society</u> 32 (2): 169-177.

    The use of consortia of microorganisms that are adapted to particular organic substrates can produce biofuel and clean the environment at the same time according to Sineriz. Extraction of fuel from biomass rather than fossil deposits may aid less developed countries and offer a solution for the world energy shortage.

Swaminathan, M. S. (1982). "Biotechnology Research and Third World Agriculture." <u>Science</u> 218 (4576): 967-972.

    See International, VI. Technology Transfer, above.

United Nations Environment Programme (1981). "Conservation/Living and Genetic Resources." <u>United</u> <u>Nations</u> <u>Environment</u> <u>Programme</u> <u>Annual</u> <u>Review</u> <u>1981</u>. Nairobi, Kenya: UNEP, pp. 57-63.

    See International, V. Organizations, above.

66

_____ (1983). "The Environmental Aspects of Energy Farms." The State of the Environment 1983. Nairobi, Kenya: UNEP. pp. 27-38.

A factual statement of how energy farms may produce fuel from biomass. Each year natural activity generates enough biomass to supply ten times the world's energy needs. Sources may be anything from crop residue to forests sugar cane or seaweed. Each has environmental pluses and minuses depending on the activities of human populations in relation to the fuel source and its ecological system. The article asks if large scale energy farming is feasible economically, socially, and environmentally.

United States Congress. House Subcommittee on Science, Research, and Technology, of the Committee on Science and Technology (1983). Environmental Implications of Genetic Engineering. 98th Congress, 1st session. Publication number 98-36. Washington, D.C.: U.S. Government Printing Office.

Transcript of a congressional hearing that considered whether deliberate release of novel organisms (plants and microbes) would damage the ecosystem. For agriculture, this includes the use of genetic biotechnology for plant growth regulation, biocontrol of pests, and frost resistance. For other areas it covers the use of engineered microbes to break down toxic waste and produce new forms of energy. The answer regarding environmental damage falls somewhere between "don't know" and "some may." These possibilities are weighed against benefits to be derived, in a context that asks what is the role of the federal government, for policy and regulation.

United States Congress, Office of Technology Assessment (1981). Impacts of Applied Genetics: Micro-Organisms, Plants, and Animals. Washington, D.C.: U.S. Government Printing Office.

A book that examines applications of 'old' and 'new' genetic technologies to microbes, plants, and animals. Chapters 1, 5, 7, 8, and 9 are of particular interest. The first is a summary chapter, and the subsequent chapters discuss the following: biomass as renewable resource; deliberate release of microbes into the environment; genetic technologies for plant agriculture; and reproductive technologies for livestock production. The book has appendices that include case studies, reproductive technologies, history of the rDNA debate, constitutional constraints on regulation, and international guidelines.

Weiss, Charles, and Nicolas Jequier, eds. (1984). Technology, Finance, and Development. Lexington, Mass.: Lexington Books.

See International, V. Organizations, above.

SECTION 5: UNIVERSITY/INDUSTRY RELATIONS AND BIOTECHNOLOGY

SECTION 5: UNIVERSITY/INDUSTRY RELATIONS IN BIOTECHNOLOGY

Ashford, Nicholas A. (1983). "A Framework for Examining the Effects
    of Industrial Funding on Academic Freedom and the Integrity of
    the University." Science, Technology, and Human Values 8 (2):
    16-23.

The purpose of this article is threefold: (1) the classification
of the types of research undertaken at the university; (2) the
examination of the concept that a university is a place of free
inquiry, diverse interests and views, and dedicated to meet
societal needs; and (3) the identification of university values
in need of preservation. The author then examines the effects of
industrial funding on these issues.

Baron, Robert (1983). "Higher Education and the Corporate Sector:
    Ethical Dilemmas." In M. Carlota Baca, and Ronald H. Stein,
    (eds.), Ethical Principles, Practices, and Problems in Higher
    Education. Springfield, Ill.: Charles C. Thomas.

In this chapter, the author discusses the evolution of the
relationship between the corporate sector and higher education,
provides a discussion of the ethical dilemmas involved, and
offers guidelines for avoiding an infringement on academic
freedom by such relationships. The author concludes that "a
well-defined, well-publicized policy of commitment to the open
dissemination of research will make plain to the corporate sector
that infringement on the university's academic freedom will not
be tolerated."

Bean, Lee Randolph (1982). "Entrepreneurial Science and the
    University." The Hastings Center Report 12 (5): 5-9.

The author is concerned with the ethical problems posed by the
increasing private capital used to fund the academic disciplines
such as biology. These ethical issues include conflict of
interest, secrecy, patents, research priorities, graduate
students, public perception, and scientific norms.

Caldart, Charles C. (1983). "Industry Investment in University
    Research." Science, Technology, and Human Values 8 (2): 24-32.

This article examines the trend of industry-funded university
research and its substantial implications for the practical,
ethical, and political areas. The author suggests some contexts
in which the issues posed in each of these areas are likely to
arise, and presents a framework for considering these issues.
The underlying premise of this analysis is "that society benefits

from the independence of the objectives of industry and
academia."

Cape, Ronald E. (1986). "Future Prospects in Biotechnology: A
Challenge to United States Leadership." In Joseph G. Perpich
(ed.), Biotechnology In Society: Private Initiatives and Public
Oversight. Oxford, Great Britain: Pergamon Press, pp. 5-9.

The industrial revolution occuring in the field of biotechnology
is the focus of this article. Cape argues that U.S. federal
support must keep pace with the R & D potential that exists
between government, universities, and the biotechnology industry
in order to maintain its national leadership role. Increased
industrial support of university research, he warns, has created
an identity crisis for the academic community. Comparisons to
the Japanese efforts are also presented.

The Center Magazine, "Part 1: New Genetics Industry Tests University
Values." 16 (3): 43-50.

This is a transcript of a discussion at a Center dialogue on the
involvement of industry in university research of genetic
engineering. Professors of several different disciplines, a
university administrator, and members of the Center Magazine
staff are among the participants that discuss the effects
university/industry relationships have on the academic community.

Culliton, Barbara (1977). "Harvard and Monsanto: The $23-Million
Alliance." Science 195 (4280): 759-763.

In this article, the author examines the unprecedented
academic-business alliance of the Monsanto Company and Harvard
University.

_____ (1981). "Biomedical Research Enters the Marketplace." The New
England Journal of Medicine 304 (20): 1195-1201.

This article examines university-industry ties in the field of
biomedical research. Examples of such relationships are
discussed. Also, concerns resulting from these relationships are
considered. These include the changes in the way research is
conducted, the effect on the university and its values, and
changes in public perception, through the press, of academic
research. The author concludes that, as government resources
decline, ". . .an industrial connection offers researchers yet
one more resource. In the interests of innovation and
intellectual diversity in science, that possibly could be
welcome."

70

_____(1982). "Pajaro Dunes: The Search for Consensus." Science 216 (4542): 155-158.

This article reports on the conference held at Pajaro Dunes which considered the implications of university-industry relationships, particularly in the field of biotechnology. An overview of the conference participants' discussion on contract disclosure, patents on licenses, and conflict of interest is presented.

_____(1982). "The Academic-Industrial Complex." Science 216 (4549): 960-962.

_____(1982). "The Hoechst Department at Mass General." Science 216 (4551): 1200-1203.

_____(1982). "Monsanto Gives Washington U. $23.5 Million." Science 216 (4552): 1295-1296.

These three brief articles discuss the recent growth of industrial investment in academic science. This series of articles examines some of the major new agreements and assesses the implications of the academic-industrial complex.

Davis, Bernard D. (1981). "Profit Sharing between Professors and the University?" The New England Journal of Medicine 304 (20): 1232-1235.

The author presents an analysis of the problems posed by the new opportunities offered through commercial interests in academic research. In this analysis the following topics are discussed: patents; private corporations and the special problems of biology; and the Harvard proposal and the future of institutional profit sharing. The author argues that there are ". . .risks in developing industrial connections, but they must be balanced against the increasing financial insecurity of universities today and against a monolithic dependence on an often unsympathetic government."

Faberge, A. C. (1982). "Open Information and Secrecy in Research." Perspectives in Biology and Medicine 25 (2): 263-278.

Faberge examines the issue of open information and secrecy in scientific research. After presenting some general remarks pertaining to this issue, including the diversity of standards with respect to secrecy, the author provides a discussion of the special cases of molecular biology and classical genetics. He concludes that open information must be consciously maintained in order to realize the benefits it offers.

71

Fowler, Donald R. (1984). "University-Industry Research Relationships." Research Management 27 (1): 35-41.

This article presents the analysis and conclusions of a survey designed to: (1) determine impediments to university-industry relationships; and (2) identify the more significant factors and determine their importance to these relationships. The questionnaire and responses are included.

"Government, Universities and Industry--The Economist Conference Papers." The Economist Intelligence Unit Special Report No. 214.

This special report describes current activities of cooperation between UK universities and polytechnics and industry. The following issues are also discussed: technology transfer from university to industry; improving university-industry communication; universities and training for industry; exploiting university expertise; licensing university inventions; and new venture formation. Appendices are also included.

Hutt, Peter Barton (1986). "University/Corporate Research Agreements." In Joseph G. Perpich (ed.), Biotechnology In Society: Private Initiatives and Public Oversight. Oxford, Great Britain: Pergamon Press, pp. 137-148.

University/corporate research agreements in the field of biotechnology is the focus of this article. The conceptual premises on which these agreements are based and the patent issues they must address are discussed. The author views such agreements as beneficial to the university, industry, and the broader public interest.

Kennedy, Donald (1982). "The Social Sponsorship of Innovation." Technology In Society 4:253-265.

The author reviews the history of federal support for biological research and the roles of government, the academic community, and industry in the innovation process. He notes the challenges posed to the government, the university, and its faculty resulting from the fundamental rearrangement of the past relationship. He then formulates a framework of proposed policy directions for the university and academic scientists in light of these changes.

_____ (1985). "Government Policies and the Cost of Doing Research." Science 227 (4686): 480-484.

Kennedy maintains that changes in the political economy of science have been the result of two trends: science is now a more capital-intensive activity while, simultaneously, federal support for research has slackened. The results of this

situation--increased funding from private industry, a greater temptation to circumvent the traditional peer review system, and increased disputes over the dispersion of available resources between institutions and investigators--has, according to the author, threatened ". . .to unravel what has been an extraordinary way of doing science."

Linnell, Robert H., ed. (1982). Dollars and Scholars: An Inquiry into the Impact of Faculty Income Upon the Function and Future of the Academy. Los Angeles: University of Southern California Press.

A large majority of academic faculty engage in work outside the university. This is the result of the nature of faculty salaries (paid on a nine month basis) and that these salaries lag behind inflation rates. The situation has fostered the potential for conflict of interest, and posed serious problems for the maintenance of academic freedom. The ethical, economic, and policy issues pertaining to the academic's outside work are discussed in this book. Summary recommendations and conclusions concerning the university's future are presented.

Linnell, Robert H. (1983). "Conflicts of Interest: Research, Consulting, and Private Practice." In M. Carlota Baca, and Ronald H. Stein (eds.), Ethical Principles, Practices, and Problems in Higher Education. Springfield, Ill.: Charles C. Thomas.

The author examines the issue of conflict of interest of academic expertise. He is particularly concerned with credibility and states ". . .if academic expertise is not credible, it is not valuable to society." These issues are discussed in relationship to society, students, and faculty. Proposals for meeting the conflict of interest challenge are presented. The author concludes, however, that "institutionalizing the academic profession" may be the best long-term solution. That is, faculty would receive salaries competitive with their counterparts in the private sector, and current professional activities practiced outside the university would be conducted, if at all, through the institution.

Mackenzie, Ian, and Roderick Rhys Jones (1985). "Universities and Industry." The Economist Intelligence Unit Special Report No. 213.

This is a collection of conference papers from The Economist conference on Government Universities, and Industry. Discussions focused on limited financial resources for research and education, and possible conflicts arising between commerical and academic objectives by private investment. Section 1 presents an international perspective on public and private research

cooperation. Section 2 discusses research relations in the UK.
A European view of investing in a competitive future is presented
in section 3. Concluding remarks are included.

Markle, Gerald E., and Stanley S. Robin (1985). "Biotechnology and
the Social Reconstruction of Molecular Biology." BioScience 35
(4): 220-226.

The authors maintain that ". . .an informed science policy must
have at its core a specification of the linkages between the
theoretical and the commerical. . . ." This article outlines
some of the theoretical and empirical relationships between
biotechnology and molecular biology. The authors first provide a
discussion on defining biotechnology. The science of molecular
biology is then examined. Biotechnology and the process of
science in relation to the dynamics that may alter the nature of
molecular biology is discussed. Finally, comments are made
concerning the future of molecular biology. Specifically, the
authors ask, "How will biotechnology affect which molecular
biology is created?" A model is presented to address this
question.

Matthews, Jana B., and Rolf Norgaard (1984). Managing the Partnership
Between Higher Education and Industry. Boulder, Colo.: National
Center for Higher Education Management Systems, Inc.

This book focuses on the issues and ideas presented at the 1983
NCHEMS National Assembly--Higher Education and Industry:
Managing the Partnership. Assessment of such partnerships, the
role of government and third parties, possible partnership
arrangements, practical managements advice, and a summary of
risks and benefits are presented. The book ". . .reflects the
broad consensus among Assembly participants that partnerships
between higher education and industry are not a necessary evil
but a new opportunity." References and a selected bibliography
are included.

Panem, Sandra (1984). The Interferon Crusade. Washington, D.C.:
Brookings Institution.

This book examines the development and promotion of interferon.
The author begins with its scientific history and events which
led to the industrial race to exploit its market potential. She
then discusses the relationship between academic and industrial
research in biological science, and resulting policy issues and
concerns. These include the constraint upon university research
as industrial investment in academic research increases, and
prospective personnel shortages in academic biology. The author
concludes with a brief discussion of policy lessons.

74

Perpich, Joseph G. (1983). "Genetic Engineering and Related
    Biotechnologies: Scientific Progress and Public Policy."
    Technology In Society 5:27-49.

    Perpich examines the NIH's impact on the development of federal
    policies in the field of biotechnology. He also provides a
    discussion on ". . .how collaboration among government,
    university, and industry might work in the light of recombinant
    DNA research initiatives in federal, state, and local
    governments—as well as some international ones." He concludes
    with a model to enhance programs among government, industries,
    and universities, particularly university/industry relationships,
    in the field of biotechnology.

Prager, Denis J., and Gilbert S. Omenn (1980). "Research, Innovation,
    and University-Industry Linkages." Science 207 (4429): 379-384.

    The status of, and potential for, formal university-industry
    cooperative relationships is the focus of this article. Types of
    university-industry relationships and barriers to enhancing these
    relationships are discussed. Furthermore, current and
    prospective federal activities for promoting these relationships
    are examined.

Science, Technology, and Human Values (1985). "Secrecy in
    University-Based Research: Who Controls? Who Tells?" Science,
    Technology, and Human Values 10(2).

    This special issue presents a number of articles by different
    authors which focus on secrecy in university-based science and
    engineering research. Part I provides a review of historical
    perspectives. Part II presents an overview of the American
    Association for the Advancement of Science's special project on
    "Openness and Secrecy in Scientific and Technical Communication."
    The focus of Part III is on handling intellectual property.
    Finally, Part IV examines national security and the universities
    in the field of computer science. Commentaries expand on issues
    raised. Also, a selected bibliography on openness and secrecy in
    science is presented.

Stark, Irwin (1983). "Industrializing Our Universities." Dissent 30
    (2): 177-182.

    The author is concerned with the rapid growth of
    university-industry relationships, and how these relationships
    affect the status of the university. He warns that to dwell on
    academic freedom issues overlooks the fundamental issue. That
    is, ". . .who is to control the university's intellectual capital
    society, to serve the public interests—or national and
    multinational corporations, to serve theirs?"

75

Stauffer, Thomas M. (1983). "Ethics of Cross-Sector Cooperation—The
Case of Business and Higher Education." In M. Carlota Baca, and
Ronald H. Stein (eds.), Ethical Principles, Practices, and
Problems in Higher Education. Springfield, Ill.: Charles C.
Thomas.

Stauffer examines cross-sectional relationships between business
and higher education in various academic fields. He discusses
the current activity of these relationships, and impediments to
interaction. He then notes several ethical debates that impede
such relationships including academic freedom, research quality,
research controls, and reserach dynamics. The realities and
opportunities of this cooperation are presented, concluding
". . .a practical basis for collaboration and the elimination
of many ethical problems does exist, only poor reasons remain for
not getting on with the task of making cooperation happen."

Thomson, Judith Jarvis, Burton S. Dreben, Eric Holtzman, and B. Robert
Kreiser (1983). "Academic Freedom and Tenure: Corporate Funding
of Academic Research." Academe 69 (6): 18a-23a.

This report, prepared by a subcommittee of the AAUP's committee
on Academic Freedom and Tenure, examines corporation funded
university research in four parts. Part I presents a brief
history of industry-university funding relationships. Part II
discusses current such relationships, and describes six different
types. Part III addresses concerns generated by these
relationships including conflict of interest, dissemination of
research results, and choice of research topic. Finally, the
recommendations of the subcommittee are presented in Part IV.

United States Congress. House Subcommittee on Investigations and
Oversight, and the Subcommittee on Science, Research, and
Technology of the Committee on Science and Technology (1982).
University/Industry Cooperation in Biotechnology. 97th Congress,
2nd session. Publication number 138. Washington, D.C.: U.S.
Government Printing Office.

This is a transcript of a congressional hearing on the
commercialization of biomedical research and, in particular,
university/industry cooperation in this field. Issues discussed
at the hearings include the following: the current status of
university/industry relationships in the field of biotechnology,
and the pros and cons of such relationships; what role the
Federal government plays in order to ensure a fair return of the
taxpayer's investment; and how industries and universities
perceive these relationships.

United States Congress, Office of Technology Assessment (1984).
Commercial Biotechnology: An International Analysis.
OTA-BA-218. Washington, D.C.: U.S. Government Printing Office.

Topics covered include the industrial use of rDNA, cell fusion,
and novel bioprocessing techniques. The competitive position of
the United States in this field with that of five other states is
assessed. Separate chapters discuss the relative importance of
ten factors that determine a state's competitive position.
Policy options are identified with respect to these factors.
Chapters 13, 14, and 17 are of special interest, which discuss
the following: government funding of basic and applied research;
personnel availability and training; and university/industry
relationships. The book includes a summary, a description of
technologies, and excellent appendices.

Wade, Nicholas (1980). "Cloning Gold Rush Turns Basic Biology into
Big Business." _Science_ 208 (4445): 688-689, 691-692.

Wade examines the implications for molecular biology as "big
business." He contends that this may result in the loss of
scientific integrity as the academic scientists link their
futures to commercial enterprises. Furthermore, he argues that
commercial interests may have also been affected by this
relationship. The author is concerned that these relationships
have created unachievable expectations.

_____(1984). "Biotechnology and Its Public." _Technology In Society_
6:17-21.

Wade discusses three key issues of biotechnology: the public
debate over hazards; the commercialization of the technology; and
the ensuing ethical issues. He then outlines the biologist's
lobbying activities against proposed regulation at the federal,
state, and local levels. He also examines the impact of the
commercialization of biotechnology on the academic biologists.
The author concludes that ". . .the more powerful biology
becomes, the more its uses and the control of those uses will be
debated."

_____(1984). _The Science Business: Report of the Twentieth Century
Fund Task Force on the Commercialization of Scientific Research._
New York: Priority Press.

This report, developed by an independent Task Force made up of
university administrators, research scientists, and corporate
executives, examines the recent wave of arrangements between
venture capitalists and research biologists at universities,
between corporations and biologists, and between corporations and
universities, in the field of molecular biology. The Task Force
concluded that these corporate-university ties are both

beneficial for the welfare of the economy and to the health of the university. In contrast, Wade, the author of the report's background paper, contends that ". . .the university's values and independence are fragile and cannot resist too much outside pressure."

Washington, Lamar, and Herbert McArthur (1983). "From Ideas to Products--Avoiding the Ethical Pitfalls." In M. Carlota Baca, and Ronald H. Stein (eds.), Ethical Principles, Practices, and Problems in Higher Education. Springfield, Ill.: Charles C. Thomas.

The authors are concerned with the ethical relationships between the academic researcher, his or her affiliated university, and the third party sponsors of his or her research project. Ethical issues pertaining to such relationships are discussed, and case histories are presented. The authors hope that the avoidance of the ethical pitfalls described will be useful for developing these "highly rewarding relationships."

Weiner, Charles (1982). "Relations of Science, Government and Industry: The Case of Recombinant DNA." In Albert H. Teich and Ray Thornton (eds.), Science, Technology, and the Issues of the Eighties: Policy Outlook. Boulder, Colo.: Westview Press.

The author reviews the background and current status of the following issues: increased university interaction with industry and its effect on the quality and direction of basic science, scientists, and the university; federally funded research and commercial exploitation; regulation; and the public's role and perception of new technology. Social and ethical considerations are also presented. The author concludes that, in this field, the focus of policy ". . . should be to help define the roles and responsibilities of scientists and the public in efforts to anticipate and shape change, rather than merely to react to it."

SECTION 6: ETHICS AND BIOTECHNOLOGY

SECTION 6: ETHICS AND BIOTECHNOLOGY

Anderson, W. French, and John C. Fletcher (1980). "Gene Therapy in
    Human Beings: When is it Ethical to Begin?" New England Journal
    of Medicine. 303 (22): 1293-1297.

    The authors summarize the state of the art of gene therapy and
    the ethical arguments that this raises.

Anfinsen, Christian (1984). "Bio-Engineering: Short-Term Optimism
    and Long-Term Risk." In Robert Esbjornson (ed.), The
    Manipulation of Life: Nobel Conference XIX. San Francisco:
    Harper & Row, pp. 42-50.

    The author argues for the human benefits of biotechnology,
    specifically through improvement of population controls and food
    resources. Furthermore, advances in biotechnology ". . .must be
    dealt with by those among us who are properly experienced in the
    moral and legal arenas of our society."

Angus, Fay (1981). "The Promise and Perils of Genetic Meddling."
    Christianity Today 25 (9): 26-29.

    In this article, the author examines genetic engineering, issues
    concerned with the technology, and provides a Christian
    perspective of these issues.

Berg, Kare, and Knut Erik Tranoy (1983). Research Ethics: Progress
    in Clinical and Biological Research, Vol. 128. New York: Alan
    R. Liss, Inc.

    In this volume, the views of various authors are presented on the
    following topics concerning research ethics: universality of
    research ethics; producers and users of scientific results;
    research, ethical standards, and responsibility; protection from
    research hazards; clinical, behavioral, and social studies
    affecting the integrity of persons; ethics at biomedical research
    frontiers; and the responsibility of science and scientists to
    provide norms for human welfare.

Blank, Robert H. (1981). The Political Implications of Human Genetic
    Technology. Boulder, Colo.: Westview Press.

    Blank presents a summary of current facts and discusses future
    prospects of biotechnology in the social and political context of
    the United States. Alternative value frameworks for dealing with
    the issues are presented. Ethical and public policy dimensions
    are examined, leading to the conclusion that these issues will
    ultimately be fought out in the political arena.

80

Bohle, Bruce, ed. (1979). Human Life: Controversies and Concerns. The Reference Shelf, Vol. 51, no. 5. New York: The H. W. Wilson Company.

This book consists of a series of articles that deal with ethical and social issues raised by abortion, artificial insemination, IVF, right to die, allocation of resources, human experimentation, behavior control, psychosurgery, organ transplantation, genetic engineering.

Capron, Alexander Morgan (1984). "Human Genetic Engineering." Technology In Society 6:23-35.

The new ethical and conceptual uncertainties, and the complex ethical and social public policy issues raised by the application of gene splicing techniques to human beings is the focus of this article. The advances in the medical uses of this technique, and their consequences are examined. The author then discusses the need for circumspection.

Cavalieri, Liebe F. (1981). The Double-Edged Helix: Science in the Real World. New York: Columbia University Press.

In this book, the internal structure of the science community is discussed and the relationship that exists between science and societal issues is examined. The author's concern is ". . .primarily with the social and often philosophic components of science and science-based technology that inevitably intersect with economic and political forces." He maintains that scientists must first consider the real significance of their work in the world picture in order to realize the contributions science has to offer.

Cole, Leonard A. (1983). Politics and the Restraint of Science. Totowa, N. J.: Rowman and Allenheld.

This book examines the debate and dilemmas concerning the restraint of sciences. Part One provides an introduction. Part Two presents historical perspectives and cases. The American political system and the restraint of science is the focus of Part Three. Turbulence, trends, and the future are discussed in Part Four, the conclusion. Discussion of these issues in context with recombinant DNA research is of particular interest.

DeWachter, Maurice A. M. (1982). "Interdisciplinary Bioethics: But Where Do We Start?" Journal of Medicine and Philosophy 7 (3): 275-287.

The first problem encountered by an interdisciplinary field is paradigm conflict. The solution offered by DeWachter is the

temporary suspension of all methods until an interdisciplinary way of stating the question is found.

Ellison, Craig W. (1979). "Engineering Humans: Who is to do What to Whom?" Christianity Today 23 (8): 14-18.

The human engineering revolution is the focus of this article. The basic issues are discussed and the value implications are examined. Specific concerns of the conservative Christian theological position are presented. The author states, "At this point, the evangelical scientist and community should be actively involved in helping to develop an ethical framework. . . ."

Fletcher, Joseph (1979). Humanhood: Essays in Biomedical Ethics. Buffalo, N. Y.: Prometheus Books.

General ethical concepts in relation to biomedical issues is the focus of this book. Fletcher makes a distinction between "rule ethics" and "situation ethics," and argues that ". . .it is wiser to be guided by moral princples than by moral rules. . . ." Chapters 7 and 16 are of particular interest which discuss genetic engineering and recombinant DNA, respectively.

Galperine, Charles, ed. (1976). Biology and the Future of Man. Proceedings of the International Conference at the Sorbonne. Paris: Universities of Paris.

Report of a conference at which these issues were discussed: organ transplantation, therapeutic experimentation on humans, genetic control of procreation, right to die, scientific responsibility. Parts of the conference were conducted in French.

Goggin, Malcolm L. (1984). "The Life Sciences and the Public: Is Science too Important to be Left to the Scientists?" With commentaries and author's response. Politics and the Life Sciences 3 (1): 28-75.

In this article and commentary package, the authors examine ". . .the arguments for expert self-rule in science and the case for more public participation in science policy making."

Gorovitz, Samuel (1984). "Against Selling Bodily Parts." QQ--Report from the Center for Philosophy and Public Policy 4 (2): 9-12.

This article presents an ethical and policy argument against the use of the free enterprise system to establish a commercial market in transplantable organs.

82

Graham, Loren R. (1981). Between Science and Values. New York:
    Columbia University Press.

    Chapter 9 is of particular interest to this section. In this
    chapter, Graham discusses ethical issues and fundamental values
    in context with biomedical technologies. He examines these
    concepts with the use of exemplary cases, including recombinant
    DNA and genetic engineering. He concludes that this technology
    ". . .often leads to serious ethical dilemmas, but it does not
    alter our fundamental values nor does it create new fundamental
    values."

Holmes, Helen B., Betty B. Hoskins, and Michael Gross, eds. (1981).
    The Custom Made Child? Clifton, N. J.: The Humana Press Inc.

    According to the editors, there are two major areas of public
    opinion that deal with value issues raised by biotechnology:
    religion and women's rights. This book consists of
    "women-centered perspectives" on value questions about
    reproductive technology, prenatal diagnosis, sex preselection,
    IVF, embryo transfer, ectogenesis.

Holtzman, Irving (1979). "Patenting Certain Forms of Life: A Moral
    Justification." The Hastings Center Report 9 (3): 9-11.

    Holtzman examines the question of whether life forms should be
    patentable. He begins by outlining the ethical issue. He then
    argues that the patenting of life forms is morally justifiable on
    utilitarian grounds.

Jonas, Hans (1985). "Ethics and Biogenetic Art." Social Research 52
    (3): 491-504.

    The author discusses the development of biogenetic techniques
    that have generated, from an ethical point of view, unprecedented
    ethical questions. He argues that previous ethical theory has
    left us unprepared to deal with these questions. He concludes
    that we must use the knowledge of our technology wisely. "But,"
    he warns, "let us not try to play creators at the roots of our
    being, at the primal seat of its mystery."

Kass, Leon R. (1985). Toward a More Natural Science: Biology and
    Human Affairs. New York: The Free Press.

    Kass presents ". . .moral and philosophical reflections on the
    powers and teachings of modern biology and medicine." He
    examines the increasing gap between science and ethics and
    explores the novel ethical issues raised by the new biology. He
    argues for a "more natural" biology, or, that which is "true to
    life as found and lived."

83

Kieffer, George H. (1979). Bioethics: A Textbook of Issues. Reading, Mass.: Addison-Wesley Publishing Company.

This a comprehensive summary and synthesis of divergent opinions on value questions raised by the "new biology." The author believes that large shifts in ethical thinking will be required to accommodate the new technology to human affairs.

Krause, Richard M. (1986). "Is the Biological Revolution a Match for the Trinity of Despair?" In Joseph G. Perpich (ed.), Biotechnology In Society: Private Initiatives and Public Oversight. Oxford, Great Britain: Pergamon Press, 31-46.

Krause discusses the biological revolution and examines how the resulting information can deal with the "trinity of despair": hunger, disease, and insufficient resources to support an expanding population. He then raises issues concerning the ". . .moral responsibility to assist the poor, the hungry, and the sick." He concludes that the biological revolution, if used wisely, can overcome the trinity of despair.

Lappe, Marc (1979). Genetic Politics: The Limits of Biological Control. New York: Simon and Schuster.

Issues discussed in this book include: the psychological consequences of genetic knowledge; the limitations and implications of the predictive power of genetic knowledge; the motives and forces that "undergird genetic research;" the relevance of genetics to epistemology; and humanizing the genetic enterprise. Lappe concludes that "Genetics is too important to be left to the geneticists."

Lygre, David A. (1979). Life Manipulation. New York: Walker and Company.

The author states that "As the biorevolution speeds ahead, we find our legal, ethical, and social values lagging behind." In this book, Lygre examines biotechnology and the ensuing social, ethical, and legal concerns. Freedom and restrictions of scientific inquiry, public safety, resources for research, and reverence for life are also discussed.

Mack, Eric (1980). "Bad Samaritanism and the Causation of Harm." Philosophy and Public Affairs 3:230-259.

This article deals with social science value issues that are relevant to organ transplantation and organ banks.

84

McCormick, Richard A. (1981). "Genetic Medicine: Notes on the Moral Literature." In his Notes on Moral Theology 1965 through 1980. Washington, D.C.: University Press of America, Inc., pp 401-422.

McCormick discusses three ethical approaches to the issue of genetic control. These are consequentialist calculus, a deontological attitude, and the mediating approach. He then presents his own personal reflections.

_____ (1981). How Brave A New World? Garden City, N.Y.: Doubleday and Company, Inc.

According to the author, policy makers need to deal intelligently with public opinion on biotechnology, of which much will be formed within the constraints of existing value systems.

McElhinney, Thomas K. (1979). "Toward the Optimal Human: Images of the Future and Genetic Engineering." In William R. Rogers and David Barnard (eds.), Nourishing the Humanistic in Medicine: Interactions with the Social Sciences. Pittsburgh: University of Pittsburgh Press, pp. 153-186.

According to the author, the ability to intervene in the human genetic structure raises value questions about optimal human functioning. Presuppositions about human nature are examined in the light of ethical arguments regarding particular genetic interventions.

Mercola, Karen E. and Martin J. Cline (1980). "The Potentials of Inserting New Genetic Information." New England Journal of Medicine. 303 (22): 1297-1300.

Organisms in nature may frequently acquire new genes from external sources both within and across species. Human insertion of genetic information can give direction to an already existing random process. This article asks when is it ethical to begin clinical trials with the new biotechnology.

Milby, T. H. (1983). "The New Biology and the Question of Personhood: Implications for Abortion." American Journal of Law and Medicine 9 (1): 31-41.

This article states that understanding the phenomena of chimerism, cloning, and parthenogenesis may relieve moral ambiguity about abortion.

Murray, Thomas H. (1985). "Ethical Issues in Genetic Engineering." Social Research 52 (3): 471-489.

Genetic engineering has attracted public attention and generated public fears. Murray discusses these technologies and groups the

ensuing ethical issues into three subsets. First, the worries about risks to public health and to the future. Second, the privatization and commercialization of genetic engineering. Finally, issues associated with gene therapy in humans. He examines these issues in context with both consequentialist and deontological ethics.

Nelson, J. Robert (1980). "Prometheus Rebounds. . .and Keeps Rebounding." In his Science and Our Troubled Conscience. Philadelphia: Fortress Press, pp. 91-150.

Religion is probably the most prominent public value system that will be applied to the new biotechnology. These chapters apply Christian ethical values to the subject matter.

Nossal, G. J. V. (1985). Reshaping Life: Key Issues in Genetic Engineering. Cambridge: Cambridge University Press.

Basic biology and genetic engineering techniques are discussed in this book. Chapter 10, "Scientists Playing God," is of particular interest to this section. This chapter examines the philosophical, moral, and ethical issues raised by genetic engineering. Issues of public policy and speculations concerning the future are discussed in subsequent chapters.

Omenn, Gilbert S. (1979). "Genetics and Epidemiology: Medical Interventions and Public Policy." Social Biology 26 (2): 117-125.

Omenn discusses the ethical, social, and public policy dimensions of public health interventions such as genetic counseling, genetic screening, and prenatal tests.

Oosthuizen, G. C., H. A. Shapiro, and S. A. Strauss, eds. (1980). Genetics and Society. Cape Town, South Africa: Oxford University Press.

This book discusses the medical, scientific, legal, sociological, religious, and ethical aspects of genetics in society. Part III, the religious and ethical aspects, is particularly relevant to this section. Viewpoints from the following groups are presented: Roman Catholic; Islamic; Judaism; Zoroastrian; Buddhist; Hindu; Anglican; and Dutch Reformed. The viewpoints range from extreme caution to welcoming the field of genetic research.

Packard, Vance (1977). The People Shapers. Boston: Little, Brown and Company.

The first two sections of this book discuss techniques for controlling behavior and for reshaping humanity. The third

section, "Concerns and Countermeasures," addresses the following: concerns about health hazards, social hazards, and individual ethics; regulation by the scientists themselves; and protection by government rules, laws, and specific administrative boards.

Pancheri, Lillian U. (1978). "Genetic Technology: Policy Decisions and Democratic Principles." In John J. Buckley, Jr. (ed.), Genetics Now: Ethical Issues in Genetic Research. Washington, D.C.: University Press of America, pp. 59-73.

Because innovations in genetic technology have the capacity to alter choice, these innovations ". . .could have the intentional or unintentional result of destroying the very form of government which initially made those innovations possible--in this country the democratic form." The author then examines policy decision-making in light of these innovations.

Powledge, Tabitha M. (1978). "Recent Social and Ethical Developments in Genetic Engineering." In John J. Buckley, Jr. (ed.), Genetics Now: Ethical Issues in Genetic Research. Washington, D.C.: University Press of America, pp. 7-23.

This paper is presented in two parts. First, the historical background of genetic engineering is discussed. The social and ethical issues concerned with this technology are then addressed. The author observes that ". . .most of the ethical and social problems that have been discussed so far have all arisen before, in other contexts."

President's Commission for the Study of Ethical Problems in Medicine and Biomedical and Behavioral Research (1982). Splicing Life: A Report on the Social and Ethical Issues of Genetic Engineering with Human Beings. Washington, D.C.: U.S. Government Printing Office.

The major ethical and social implications and concerns of genetic engineering are the focus of this report. The meaning of the term "genetic engineering," concerns about the technology, and the objectives and process of the study is discussed. Also, a description of some of the most important techniques of the technology is provided. The ensuing social and ethical issues are examined and concerns for the future are addressed. Appendices and a summary of conclusions and recommendations are included.

Shannon, Thomas A. (1981). "A New Design for Life: Ethics and the Genetic Revolution." New Catholic World 224 (1341): 111-113.

In this article, the author identifies and describes several of the emerging ethical issues concerned with the genetic revolution. He states, "These issues are important not only

because how we resolve these issues will say a lot about ourselves and our society but also for the very simple and pressing reason that we can no longer avoid these issues."

Wall, Thomas F. (1980). <u>Medical Ethics: Basic Moral Issues</u>. Washington, D.C.: University Press of America.

This book provides a general introduction to the ethical problems posed by recent developments in medical research, practice, and technology. Wall begins with a discussion of ethical theory and of the criteria for moral decision-making. He then presents the following issues: genetic engineering; behavior control; euthanasia; abortion; human experimentation; and health care. Various proposed solutions are examined with respect to these issues.

Wade, Nicholas (1984). "Biotechnology and Its Public." <u>Technology In Society</u> 6:17-21.

Wade discusses three key issues of biotechnology: the public debate over hazards; the commercialization of the technology; and the ensuing ethical issues.

Wall, Thomas F. (1980). <u>Medical Ethics: Basic Moral Issues</u>. Washington, D.C.: University Press of America.

This book provides a general introduction to the ethical problems posed by recent developments in medical research, practice, and technology. Wall begins with a discussion of ethical theory and of the criteria for moral decision-making. He then presents the following issues: genetic engineering; behavior control; euthanasia; abortion; human experimentation; and health care. Various proposed solutions are examined with respect to these issues.

Weiner, Charles (1982). "Relations of Science, Government, and Industry: The Case of Recombinant DNA." In Albert H. Teich and Ray Thornton (eds.), <u>Science, Technology, and the Issues of the Eighties: Policy Outlook</u>. Boulder, Colo.: Westview Press, pp. 79-97.

The author reviews the background and current status of the following issues: increased university interaction with industry; regulation; and the public's role and perception of the new technology. He then examines the social and ethical consequences posed by this technology. In order to "restore communication and confidence," Weiner maintains scientists and nonscientists should explore these ethical issues together.

88

Zimmerman, Burke K. (1984). <u>Biofuture</u>: <u>Confronting the Genetic Era</u>.
New York: Plenum Press.

In this book, Zimmerman explains the intricacies of biotechnology
and cell biology. He states, "With that knowledge comes the
responsibility to use it wisely." The power this knowledge has
given to man and the ethical dilemmas it raises are then
discussed. The author examines these issues in relation to past
and present and presents his own perceptions concerning the
future.